Interdisciplinary Insights from the Plague of Cyprian

Mark Orsag · Amanda E. McKinney ·
DeeAnn M. Reeder

Interdisciplinary Insights from the Plague of Cyprian

Pathology, Epidemiology, Ecology and History

palgrave
macmillan

Mark Orsag
Department of History
Doane University
Crete, NE, USA

Amanda E. McKinney
Institute for Human
and Planetary Health
Crete, NE, USA

DeeAnn M. Reeder
Bucknell University
Lewisburg, PA, USA

ISBN 978-3-031-26093-3 ISBN 978-3-031-26094-0 (eBook)
https://doi.org/10.1007/978-3-031-26094-0

Cover illustration: Maram_shutterstock.com

This Palgrave Macmillan imprint is published by the registered company Springer Nature Switzerland AG
The registered company address is: Gewerbestrasse 11, 6330 Cham, Switzerland

Acknowledgments

Firstly, I would like to thank Bucknell University, my academic home for the past 17 years, for its unflagging support of my scholarship and my professional development. From my colleagues to my students, Bucknell has nurtured my sense of curiosity and wonder and allowed me to pursue myriad interests. This book is the culmination of one such interdisciplinary adventure. I am so grateful to Director Brian Pope of the Lubee Bat Conservancy in Gainesville, Florida for introducing me to my co-authors Dr. Amanda E. McKinney and Dr. Mark Orsag from Doane University. The Institute for Human and Planetary Health (IHPH) is an extraordinary and forward-thinking embodiment of the One Health perspective. It has been my privilege to partner with them on this project.

My 20+ years of research on bats has been made possible through great mentorship and collaborations. Dr. Thomas Kunz first introduced me to bats and opened the door to explore their extraordinary secrets. He was a giant among the bat research and conservation community and is missed by all. Although I studied primates for my Ph.D., the exceptional mentorship provided by Dr. Sally Mendoza and Dr. William Mason made me the scientist that I am today. My studies in Africa have been enhanced by my partnership with the Ugandan scientist Imran Ejotre—whose perspectives and friendship are both highly valued. Drs. Jonathan Towner, Brian Amman, Serena Carroll, and Brian Bird provided my introduction to filoviruses and provide support and encouragement to this day. Lastly, Dr. Kenneth Field, also of Bucknell University, has been my long standing (if not sometimes reluctant) "card carrying immunologist" friend and collaborator—my studies of bat disease ecology would not be successful without him.

On a personal note, all that I do is possible because of the love and support of my husband of 32 years, Thomas Reeder. Together we have roamed the world, raised two children (who benefited greatly from their multiple trips to Africa, except for the malaria part), completed five degrees, moved fifteen times, had a farm, loved six dogs, and survived a pandemic. Our daughter, Sophia, is now grown and has a PhD of her own. Not surprisingly, she studies malaria. We couldn't be prouder.

Dr. DeeAnn M. Reeder

First, and foremost, I'd like to thank my co-authors, Dr. DeeAnn Reeder, and most importantly, Dr. Mark Orsag, without whom this book would not have been written. I'd also like to thank our students, whose ability to open their eyes and see the realities hiding in plain sight inspires me to contribute whatever I can to help make their futures brighter. I am also indebted to Dr. Jacque Carter, former President of Doane University. Without Jacque's foresight, insight, friendship, and counsel, the Institute for Human and Planetary Health would never have been born. He fostered an environment where interdisciplinarity could grow, leading to, among many other things, this very project. I'm also grateful for Melissa Clouse's editing abilities and, even more, her friendship.

Last, but certainly not least, I'd like to thank my family. My husband, Patrick, is the rock upon which I stand, and who makes it possible for me to pursue all my (sometimes crazy) passions. Without his devotion to me and our children, our family unit would not function! My children, Callan and Aidan, are my heart and my joy. They drive me to do (hopefully) meaningful work so that the planet they inherit might eventually benefit, but more importantly, that they will be inspired to seek to do meaningful work in the future themselves. I'd also like to thank my parents for bringing me into the world, relentlessly fostering my curiosity, modeling empathy and generosity in all things, and for always supporting me, my family, and my pursuits.

Dr. Amanda E. McKinney, M.D.

Institutionally, I would like to thank the Institute for Human and Planetary Health (IHPH), in particular, for inspiring and supporting this project. We are all also grateful to Bucknell University for its generous support of this project in multiple ways. Also, thanks to both Doane University's Tech Service Center and Kimm Dilocker of Script Quill n'Type for their able help and assistance. I also owe so much to my amazing, beloved, and perpetually insightful wife Rebecca (and our funny friendly Welsh terrier, Dylan). Thanks as well to my highly erudite mother, Ann and brother Gregg for everything! My gratitude goes out also to Gregg's family (Ellen, Ty, and Ian), and also to the Goff (Dan, Nan, and Campbell), Steger (Gary, Kim, Taylor and Jordan), and Word (Scott, Brenda, Zack, Sara, Ellie, Brooks, and Piper) families. Special thanks to my Mother-in-law Norma whose decency and kindness are profound indeed.

Thanks to my two amazing, wonderful, and invaluable "tripod" co-authors, Dr. Amanda E. McKinney and Dr. DeeAnn M. Reeder. I extend thanks also to Jack Sutton of Bucknell University for his highly skilled technical help. Also, I am grateful to both Lucy Kidwell, our very insightful and helpful History Editor at Palgrave Macmillan and to the precise Noorjahan Begum, our Production Editor at Springer Nature, as well as to the highly organized and helpful Mathru Srinivasan Vaitheeshwaranb (at Straive). Profound thanks also go out to all my helpful friends and colleagues (present and former) at Doane University—including the unforgettable Dr. Andrea Holmes, Andrea Butler, Melissa Clouse, Dr. Jay Gilbert, Dr. Brandi Hilton-Hagemann, Derek Biermann, Dr. Liam Purdon, Dr. Kurt Runestad, and Joel Weyand. Special personal thanks to former Doane University President Dr. Jacque Carter for all his steadfast support of my work over the years. I also owe a deep debt to my tireless colleague Dr. Kim Jarvis. Gratitude is also owed to Dr. Kyle Harper; his truly pioneering scholarship on the Plague of Cyprian inspired, shaped, enabled, and guided our own efforts in so many ways. Without his knowledge and insight, this work would not exist. Also, thanks to Dr. Sabine Huebner, whose important more recent scholarship on the topic displayed immense erudition in terms of utilizing ancient source materials and also helped validate, shape, and better articulate some of our conclusions. Despite some ultimately differing conclusions, her scholarship prompted us to make important analytical corrections and guided us in filling key evidentiary gaps in ways that we never could have without the benefit of her work.

I also thank my friends and rock-climbing partners, the Ortons (Alaina, Brian, and last but certainly not least Ev), as well as Lindsay, Rob,

Cheyenne, Matt, and Wendy, you helped keep me sane through this demanding project and the unfolding pandemic. Finally, a special thank you to my mentor at Michigan State University Dr. Lewis Siegelbaum from whom I directly learned so much and to whom I owe so much of what I have since learned. Finally, to those who are no longer here, but who are revered in memory: my father Harry and my Father-in-law Zack, (along with our amazing Welsh terrier companion of 12 years, Griffin). In closing, I would like to note the significant contributions to this project of the following:

Melissa Clouse, *MS*. Faculty, Health Sciences, Colorado State University-Global
Lindsay Sears, *MS*. Interactive Animal Programs Manager, Omaha's Henry Doorly Zoo and Aquarium
Caitie Welty, Documentation Specialist, Doane University Center for Computing in the Liberal Arts

Dr. Mark Orsag

Contents

About the Authors

Dr. Mark Orsag, Ph.D. is Professor of European and Interdisciplinary History at Doane University. He is also an Advisory Board Member in IHPH. Dr. Orsag received his B.A. from Carnegie-Mellon University, his M.A. from the Pennsylvania State University and his Ph.D. from Michigan State University. He is the current Chair of Doane University's History Department. His areas of concentration in teaching, professional presentation, and published scholarly output have been wide-ranging and encompassed Ancient, European, Genealogical, Russian, Military, and Interdisciplinary history. As a teacher, Dr. Orsag has received three Doane University Student Congress Teaching Awards. His current scholarly interests are collaborative projects centered at the nexus of history and the natural sciences. This includes work with Doane University colleagues Dr. Andrea Holmes and Dr. Amanda E. McKinney, M.D., as well as with Dr. DeeAnn M. Reeder, Ph.D. (Biology) of Bucknell University. The latter two being the co-authors of the five sections that originally collectively comprised the Plague of Cyprian project. All of these were accepted and presented at the 2021 and 2022 Association of Ancient Historians Conference (University of Illinois and University of California-San Diego). All five were also accepted and presented at the American Historical Association Conference in 2022. Dr. Orsag's other recent scholarly projects include an in-progress interdisciplinary monograph (also with Dr. McKinney), *The 21st First Century Crisis: Healing in the Anthropocene*. He was also a multi-chapter lead or second author and served as the Editor for *Cannabis: A Comprehensive Overview* (Bowker/Amazon, 2021), a recently published, multidisciplinary/multi-author two volume textbook set written under the leadership of Doane University Professor of Chemistry and Director of Cannabis Studies, Dr. Andrea Holmes. Holmes' research has been repeatedly nationally awarded; she is a recipient of the National Research Service Award (from the National Institutes of Health), the National Science Foundation's Presidential Career and Research Infrastructure Improvement Awards, and the Henry Dreyfus Teacher Scholar Award.

Dr. Amanda E. McKinney, M.D., CPE, FACLM, FACOG is the Founder and Executive Director of the Institute for Human and Planetary Health (IHPH) in the USA. She is a triple-board certified physician with a medical degree from the University of Nebraska and residency/fellowship training from the University of California-Irvine. She is a Collaborator in the Planetary Limits Academic Network (PLAN), "which aims to raise awareness about critical systemic challenges facing the human endeavor." Her ongoing research encompasses both plant medicine and how planetary limits will impact US healthcare. Dr. McKinney has given over twenty invited presentations, at times as a keynote or plenary speaker, in the United States and overseas. She has published articles in, among other publications, *The Lancet Planetary Health, The American Journal of Lifestyle Medicine* and the *American Journal of Obstetrics and Gynecology* along with multiple textbook chapters on women's health, lifestyle medicine, and medical cannabis. She was also featured, along with Samuel L. Jackson, James Cameron, and others, in the 2016 documentary film *Eating You Alive*. In 2021, she was featured in a second documentary film *PlantWise*.

Dr. DeeAnn M. Reeder, Ph.D., Professor of Biology at Bucknell University, is a wildlife biologist who studies disease ecology, behavior, physiology, and conservation. She holds a B.A. in Zoology from UC Berkeley, M.S. and Ph.D. degrees in animal behavior from UC Davis, and completed postdoctoral research at Boston University. Her current research explores the relationships between bat health, ecosystem health, and human disease risk. She is recognized as an expert in Pteropodid (fruit) bat physiology and behavior and has established field sites in Uganda and South Sudan. In her work, she seeks to quantify the response of free-ranging African bats to filoviruses in order to test the hypothesis that viral tolerance by these reservoir hosts is mediated through specific adaptations in immune signaling pathways. Dr. Reeder has also contributed significantly to our understanding of the deadly white-nose syndrome (WNS), which has killed millions of bats in North America in the past decade. Dr. Reeder holds a research position at the National Museum of Natural History, Smithsonian Institution in Washington DC. In addition to her bat research, she is recognized internationally for her studies of mammal biodiversity, especially in South Sudan. Her commitment to the conservation of global mammal biodiversity is evident in her editorship and management of the Mammal Species of the World project. Her work has been funded by a number of agencies, including National Institutes of Health, National Science Foundation, National Geographic, US Fish & Wildlife Service, and USAID. In 2021, Dr. Reeder was awarded a $3 million National Institutes of Health grant related to filovirus research in Uganda.

Abbreviations

AHA	American Historical Association
ALIMA	Alliance for International Medical Action
BDBV	Bundibugyo Virus
BOMV	Bombali Virus
BSL-4	Biosafety Level-4
CAFOs	Confined Animal Feeding Operations
CCHF	Crimean-Congo Hemorrhagic Fever
CDC	Centers for Disease Control
CFR	Case Fatality Rate
DNA	Deoxyribonucleic Acid
DRC	Democratic Republic of the Congo
EBOV	Ebola Virus
ECDC	European Center for Disease Prevention and Control
ELISA	Enzyme-Linked Immunoassay
ERB	Egyptian Rousette Bat
EVI	Enhanced Vegetation Index
GARP	Genetic Algorithm for Rule-Set Production
GBIF	Global Biodiversity Information Facility
GP	Glycoprotein
HA	Hectare
HFRS	Hantavirus Hemorrhagic Fever with Renal Syndrome
HPD	Highest Posterior Density (Mathematics)
IHPH	Institute for Human and Planetary Health
IUCN	International Union for Conservation of Nature
LBRF	Louse-Borne Relapsing Fever
LLOV	Lloviu Virus

MARV	Marburg Virus
MERS	Middle East Respiratory Syndrome
MLAV	Mengla Virus
MRCA	Most Recent Common Ancestor
MVD	Marburg Virus Disease
NASA	National Aeronautics and Space Administration
NCDs	Non-Communicable Diseases
NDVI	Normalized Difference Vegetation Index
NERVEs	Non-Retroviral Endogenous RNA Viral Elements
NIH	National Institutes of Health
NS/S/Y	Nucleotide Substitutions/Site/Year
ORA	Oxford University Research Archive
PCR	Polymerase Chain Reaction
POC	Plague of Cyprian
PPE	Personal Protective Equipment
PPV	Posterior Probability Values (Mathematics)
R_0	In epidemiology, the basic reproduction number of a particular pathogen in a given environment
RAVV	Ravn Virus
RCO	Roman Climate Optimum
RESTV	Reston Virus
RNA	Ribonucleic Acid
RVF	Rift Valley fever
SARS	Severe Acute Respiratory Syndrome
SES	Social-Ecological System
SUDV	Sudan Virus
TAFV	Tai Forest Virus
TBRF	Tick-Borne Relapsing Fever
T_g	Epidemiological measure of time between infectiousness in primary and secondary cases of a disease
TMRCA	Time to Most Recent Common Ancestor
TOW/DS	Total Outbreak Window/Dissemination Score
VHF	Viral Hemorrhagic Fever
WGS	Whole Genome Sequencing
WHO	World Health Organization

List of Maps, Figures, Images, Graphs and Diagrams

List of Tables

CHAPTER 1

Introduction

> It's not the destination, it's the journey.—Ralph Waldo Emerson, *Self-Reliance*, 1841

Our project had several interconnected and complementary wellsprings and later evolved to have four main purposes. More narrowly, it was born from Amanda McKinney's Institute for Human and Planetary Health (IHPH) and its missions of advocacy and education regarding the strongly interdisciplinary, complex, and crucially current mission of understanding the intricate linkages between the health of the planet and human viability. This is a perspective that strongly informed this project. These are interconnections related to diet and food systems, climate, disease, medicine, and strategies for sustainability and resilience in response to a multiplicity of twenty-first-century global challenges. It was the product of DeeAnn M. Reeder's relentless quest to achieve greater understanding of the ecology of globally menacing diseases such as Ebola. It was the indirect result of Dr. Mark Orsag's fascination with interdisciplinary history and search for informative parallels within it. Firstly, this project was also born of the authors' shared belief in the importance of interdisciplinary "systems-synthesis"-based research/analytical methods. We felt these were oft-advocated but far more rarely realized in modern higher educational environments. More disciplinarily defined, established approaches undoubtedly work superbly for many academic research problems. After all, they have become modern conventions for a reason.

M. Orsag et al., *Interdisciplinary Insights from the Plague of Cyprian*,
https://doi.org/10.1007/978-3-031-26094-0_1

Yet, these approaches are, by their very nature, often narrow in scope. Our contrasting aim is to build broader, more holistic, and mutually supporting interdisciplinary consilience when tackling the considerable challenge of analyzing a time-obscured complex pandemic which, if looked at from a narrower disciplinary perspective, has many evidentiary pieces missing or open to multiple defensible interpretations. Our overall approach to our topic is, in certain ways, quite novel and will likely engender some controversy. Such approaches do indeed face potential pitfalls of circular or reductionist reasoning. We have striven throughout to remain cognizant of and to avoid, analytically, those problems.

Secondly, we explicitly approached the Plague of Cyprian as a challenging multidisciplinary puzzle to be "solved" to as high a degree of certainty as possible. A convincing retrospective diagnosis in terms of the likely pathogen involved is indeed a key component of it. Though our topic is largely historical, we utilized analytical methods drawn not only from the discipline of history but also from very different ones—such as modeling and other techniques derived and extrapolated from the experimentally based approaches of the natural sciences. We believe that such systems-synthesis-based methodologies could also be important both in gaining greater scholarly understanding of the past but also in helping confront threatening current and interconnected global problems. If our work, in some small way, can help promote wider scholarly use of such explicitly interdisciplinary approaches, that would be a great reward in and of itself. Additionally, for reasons that we will detail, the Plague of Cyprian is also a heinously difficult puzzle unlikely ever to be pieced together or "solved," more narrowly, by history or science alone. An interdisciplinary approach to analyzing this ancient pandemic, producing a "solution" potentially based in interdisciplinary probability as opposed to scientific certainty, seems to us, the best way forward.

As we came to more fully realize as we progressed, the core of our work and methodology fell largely within the confines of the "consilient approach to history" associated with such eminent scholars as Dr. Michael McCormick of Harvard University. As McCormick put it in 2011:

> Consilience refers to the quality of investigations that draw conclusions from forms of evidence that are epistemologically distinct. The term seems particularly apt for conclusions produced by natural-scientific investigations on the one hand and by historical and archaeological studies on the other.

> Consilience points to areas of underlying unity of humanistic and scientific investigation– a unity arising from that of reality itself; it represents a convergence in parallel but independent investigations that results in deductions that are much more robust than any investigation would be able to produce on its own.

Temporally, the origins of our work are situated in the late winter/early spring of 2020 with an interdisciplinary course entitled *Apocalypse: How Societies Survive and Fail to Survive Existential Threats*, co-taught by Amanda and myself. Even as the class began, the early stages of the Covid-19 pandemic were well under way. Our class was structured around, among other topics, the historical and ongoing threat from zoonotic viral pandemics. Soon, the whole world found itself in the middle of one. One of the books that we used, during the period of time that the thirty-four students and two professors were still meeting in person (not yet having been forced onto the virtual world of necessity called *Zoom*), was Dr. Kyle Harper's (Harper had studied under and published with Dr. McCormick at Harvard) *The Fate of Rome*. We were intrigued and impressed by the work's impressive scope, interdisciplinary approaches and bold conclusions. The book also seemed to embody so much about, albeit in an ancient historical context, the critically important (even more so in light of the ongoing pandemic) nexus of ecology, disease and human society that IHPH was dedicated to exploring. We were particularly struck by Harper's impressive marshaling and interpreting of obscure, patchy and open to interpretation evidence and yet coming to what seemed like highly defensible conclusions regarding this "forgotten pandemic"—the mysterious yet vividly described third-century Plague of Cyprian. At first, however, Harper's implication of a filovirus as the pathogenic cause of the pestilence seemed odd to us—a jarring and improbable blending of the ancient and modern, as well as of the biogeographies of the Mediterranean and Sub-Saharan Africa. As we thought more about the relevant issues, we became ever more intrigued and decided to do "a little research" on our own. Our initial aim was to publish a "forty-page" article narrowly agreeing or disagreeing with Harper's conclusions. After all, we had the generally right blend of interdisciplinary skills—a physician and a historian, both with knowledge of and interest in these precise types of issues.

Soon, however, the project began to grow far past our original conception of it. As the Covid-19 pandemic around us engulfed a terrified and baffled world, we began to realize that we didn't understand everything that we needed to "solve" the heinously complicated puzzle of this third-century pandemic on our own. Amanda and I reached out for help and found the exact right person, Dr. DeeAnn M. Reeder of Bucknell University. DeeAnn graciously signed on and brought her invaluable wealth of knowledge and experience to the project. From then on, it was impossible to conceive of this work successfully "standing" without each author in our "tripod." Three other contributors, our Doane University friend and colleague Melissa Clouse (Program Director Health Sciences), Lindsay Sears of Omaha's Henry Doorly Zoo and Aquarium, and then Doane University student (and liberal arts computing specialist) Catie Welty, supplemented the work of the three main authors.

While our rather diverse group of scholars (whose academic skill-sets ranged from that of a more generalist historian to that of a truly world class biologist) worked, perhaps surprisingly, very well and cooperatively together, we did encounter numerous instances that demonstrated why interdisciplinary research in the discipline-defined world of modern higher education has a number of inherent difficulties. We frequently used digital sources and methodologies. Due to operating in spaces that effectively fell between, or were relatively untreated within, traditional disciplinary boundaries, we occasionally were forced to rely on government or commercial sources as opposed to more purely academic ones. Such, even limited, reliance created its own set of complications. We also encountered fundamental differences in research and analytical approaches and methods. Certain words had completely different meanings in various disciplinary contexts. Differing citation and reference conventions also became apparent. We also encountered, at times, mutual incomprehension whether in matters of terminology or more fundamentally in terms of knowledge. We patiently worked to educate each other. This usually consisted of teaching, as well as correction, in the fields of science and medicine for the generalist historian. This sort of education, however, also allowed the generalist, at times, to act as a kind of translator between the two more erudite scientists. Hopefully, if such interdisciplinary team-based approaches are applied more consistently, there can be continuing refinement in terms of working toward ever more specifically knowledgeable and effective multidisciplinary team compositions.

In analytically linking details from the disciplines of history, archeology, numismatics, wildlife biology, ecology, virology, medicine, molecular phylogenetics, etc., we also often found ourselves connecting the ancient Roman and modern worlds in ways that seemed to demonstrate the intense relevance of the former for the latter. We were exploring complex interdisciplinary connections among Mediterraneanization, animals, ecosystems, anthropogenesis, patterns of human movement, climate, social practices and institutions, medicine (including both epidemiology and pathology), and disease in ways that seemed particularly timeless and highly relevant. The world of the Plague of Cyprian and the Roman Third-Century Crisis indeed had, we believed, relevance for the globalized, destabilized, and fossil-fueled third decade of the twenty-first century.

The other main work that Amanda and I used in that foundational 2020 class, back at the Covid pandemic's beginning that now seems a different world entirely, was Ugo Bardi's *The Seneca Effect: Why Growth Is Slow but Collapse Is Rapid* [Springer, 2017]. This is an impressive and thought-provoking work of complex systems theory devoted to examining, in myriad ways, the mechanics of the collapse of complex systems as a phenomenon. Or, as the book's preface more memorably puts it, analyzing collapse as "a feature of the universe." Like the Plague of Cyprian in the Roman third century or, alternatively, the "forcing generated by the depletion of precious metal mines" that Bardi implicates in the Western Roman Empire's fifth-century "Fall," the Covid-19 pandemic seems to have put the problems of the twenty-first century's globalized world "...in high relief." Bardi further notes applicably that "...a complex system may multiply the effect of the [external] perturbation many times, as when you scratch a match against a rough surface." Such amplification can bring destructive collapse and entropy. Alternatively, it can (also conversely through dampening as opposed to amplification) promote adaptation, transformation, and resilience in response to collapse; or result in some complex combination of the two. Kyle Harper seems, quite eloquently, to embody the former interpretation/outcome in the following passage from *The Fate of Rome: Climate Disease & the End of an Empire* [Princeton, 2017]:

> The story of Rome's end is a human one. There were tense moments when human action decided the margin between triumph and defeat. And there were deeper, material dynamics-- of agrarian production and tax collection, demographic struggle and social evolution-- that determined the scope and success of Rome's power...The Romans built a giant Mediterranean empire

> at a particular moment… a moment suspended on the edge of tremendous natural climate change. Even more consequentially, the Romans built an interconnected, urbanized empire on the fringes of the tropics, with tendrils creeping across the known world. In an unintended conspiracy with nature, the Romans created a disease ecology that unleashed the latent power of pathogen evolution [and zoonosis]. The Romans were soon engulfed by the overwhelming force of what we would today call emerging infectious diseases. The end of Rome's empire, then, is a story in which humanity and environment cannot be separated. Or, rather, it is one chapter in the still unfolding story of our relationship with the environment. The fate of Rome might serve to remind us that nature is cunning and capricious. The deep power of evolution can change the world in a mere moment. Surprise and paradox lurk in the heart of progress.

Yet other eminent scholars contrastingly see a less clearly defined Roman collapse and more of a complex *longue durée* ambiguous continuation, survival, transformation, and adaptation. The following revelatory quotation is from Dr. Michelle Salzman's (University of California-Riverside) *The Falls of Rome: Crises, Resilience and Resurgence in Late Antiquity* [Cambridge, 2021]:

> If we view the world as the Romans at the time did and consider how individuals and groups reacted to these [Salzman had, above, discussed events surrounding the "ruinous" Vandal Sack of Rome in 455 CE] and other events that they themselves regarded as crises, we can see that senators, emperors, bishops and generals also interpreted these events as opportunities to advance their own positions or viewpoints. Roman elites in these centuries demonstrated what social scientists call *resilience,* defined as the marshaling of resources to reorganize and restore social formations in the face of fractures and swerves. Although social scientists have developed this model to analyze environmental shocks on societies or to consider state-level interventions to mitigate the consequences of catastrophic events like plagues or earthquakes, I use the term to consider how Roman elites adapted to the shocks from political and military crises that overtook the city of Rome during the last three centuries of its existence. Thus, I follow scholars who study how the "resilience of a society affects other groups and institutions within the same society" and acknowledge that the burden of recovery and its costs are not shared equally.

Thirdly and fourthly, what indeed are the lessons from the unraveling or alternatively transformation of the ancient Mediterraneanized Roman world for the twenty-first-century globalized one? There are indeed no simple answers. Could we, however, more fully illuminate, through use of an interdisciplinary "systems-synthesis" approach, the probable cause and mechanics of a single piece of that much vaster puzzle—a significant perturbation of Roman civilization by, in Harper's words, the "elemental forces" of disease and nature and their interaction with imperial society in the third century in terms of the Plague of Cyprian's pathology, epidemiology, ecology, and history? Could we begin to understand how Roman civilization was affected and then adapted and showed resilience (or didn't)? Would that help us gain insight and sustain amidst the challenges (both similar and different) facing us in the twenty-first century? We have also written, in part, in that latter hope.

PART I

"The Theory": The Rediscovery and Reinterpretation of an Ancient Pandemic

"If I have seen further, it is by standing on the shoulders of Giants."—Sir Isaac Newton, 1675.

CHAPTER 2

The Ancient Evidentiary Foundations

In the crisis-ridden mid-third century, an unfamiliar and horrific pandemic disease stalked the Roman Empire. The illness is best known through the detailed, vivid, and remarkable account of its pathology and devastating broader impact by prominent early Christian Church figure St. Cyprian of Carthage; so much so that his name was eventually attached to the disease event. Thascius Caecilius Cyprianus was born around 200 CE and was martyred in 258 CE during a persecution of Christians triggered by the Two Edicts of the Roman Emperor Valerian (r. 253–260). Cyprian was an attorney by training and a skilled rhetorician. As noted by Dr. Barbara Logan of the University of Wyoming, in her paper "Plague, Persecution and Purpose in St. Cyprian's *De Mortalite*" (AHA 2022), Cyprian was famed in the early church for his faith and eloquence. St. Cyprian was described by the fourth–fifth-century Christian poet Prudentius as follows. "...God's grace so expands itself in the exuberance and richness of his discourse that he will never cease to speak, even to the end of the world...". While much remains mysterious about the plague that Cyprian described in detail, the terrifying extreme lethality of the hope-crushing and dread-inducing pathogen involved seems clear, even over a distance of over 1,700 years, from his striking account *De Mortalite*.

M. Orsag et al., *Interdisciplinary Insights from the Plague of Cyprian*, https://doi.org/10.1007/978-3-031-26094-0_2

In a series of articles and opinion pieces, as well as in his groundbreaking historical monograph *The Fate of Rome: Climate, Disease, and the End of an Empire* (Princeton University Press, 2017), prominent classicist Dr. Kyle Harper hypothesized that this apparently unique and highly destructive pestilence that ravaged the Roman Empire was probably caused by an outbreak of one of the members of the deadly *filoviridae* (a family of diseases that encompasses the modern Ebola and Marburg viruses—two of the deadliest pathogens on Earth). It is notable that the appellation "The Plague of Cyprian" is very modern in origin. The Romans certainly did not refer to the disease in that way, and specific terms such as "The Great Pestilence," which the famed Greco-Roman physician Galen applied to the second-century Antonine Plague, are missing from surviving ancient accounts of this mid-third-century pandemic. It was perhaps, from an ancient Roman perspective, a nameless, unstoppable, unfamiliar, and unfathomable dread.

Harper's filovirus hypothesis disputed the rather limited existing historiographical consensus initially developed by historians such as Dr. William Hardy McNeill of the University of Chicago and Dr. Dionysios Stathakopoulos of the University of Cyprus. They had blamed the Plague of Cyprian on the aerosolized pathogen smallpox; or, in McNeill's case, perhaps, alternatively, on the also airborne and highly infectious disease measles. In his classic 1976 monograph *Plagues and Peoples*, McNeill noted Cyprian's account of the disease (but still did not use the appellation "The Plague of Cyprian") and the "intrinsically persuasive" if "unproven" possibility that the plague outbreak had contributed to the rise of Christianity in late antiquity. McNeill also believed the mid-third-century pandemic to be "fully comparable" in its impact to the earlier Antonine Plague. He notes high levels of mortality from the pestilence in both urban (5,000 deaths a day in Rome at one point) and rural settings; this was, in McNeill's view, a terrible and truly universal pestilence. Harper's groundbreaking work provoked some renewed scholarly interest (including our own) in the topic. In the summer of 2021, the noted classicist Dr. Sabine Huebner of the University of Basel published a lengthy and detailed article, "The 'Plague of Cyprian': A Revised View of the Origin and Spread of a Third Century, CE Pandemic" in the *Journal of Roman Archaeology*. This work offered a significant and highly developed critical and contrasting counterpart/alternative theory of the pestilential event to

Harper's re-conceptualization of the Plague of Cyprian. Yet, despite her questioning of a number of Harper's conclusions regarding the Plague of Cyprian, Huebner noted "Harper's very useful series of articles on this epidemic, which constitute its first comprehensive treatment and filled a real lacuna in our knowledge about ancient epidemics" [21, p. 152; 27, pp. 104–108].

Harper's research and analysis, while grounded in a close and historically expert reading of various ancient sources, was also somewhat interdisciplinary in that the author drew upon twenty-first-century medical, molecular-phylogenetic, and climate science knowledge as well as numismatic and archaeological evidence to reach his conclusions. Huebner's later article also drew upon a wide range of interdisciplinary evidence in examining the Plague of Cyprian. Harper's entire corpus of scholarship regarding the Plague of Cyprian—his later (2017 onward) academic articles, the Plague of Cyprian-focused section of his broader scholarly monograph and other more popularly oriented works—also build upon detailed evidentiary and source analyses contained in a series of articles that he published in 2015–2016 in the *Journal of Roman Archeology* and the *Journal of Economic History*. At the heart of his extensive research (and very important as well as to Dr. Huebner's), are a variety of contemporary early Christian sources, including the eyewitness testimony of Bishop Dionysius of Alexandria (whose work survives today through the writings of fourth-century church chronicler Eusebius Pamphilus) and the thoroughly remarkable aforementioned account of Cyprian, Bishop of Carthage. Dr. Harper, holistically, makes a convincing initial case that a filovirus was potentially responsible for a deadly pandemic that was a contributing factor in the Roman Empire's Third-Century Crisis. Filoviruses, along with a handful of other virus groups (e.g., flaviviruses, arenaviruses, and hantaviruses) make up the broader group of diseases collectively known as viral hemorrhagic fevers (VHFs). The factors behind this consensus-challenging conclusion gleaned from a combination of Harper's reading of the ancient sources and differential analysis of the evidence concerning the pathology and epidemiology of the Plague of Cyprian, are conveniently summarized in *The Fate of Rome's* Table 4.1 [16, p. 144]. Its overall results indeed seem to strongly indicate that the likely pathogen that caused the Plague of Cyprian was a VHF. We have created our own broader list of thirty elements of pathology and epidemiology—drawn from both Harper's work and our own research:

List 2.1 Defining the Plague of Cyprian, Medically: Thirty Elements of Pathology and Epidemiology Created by Dr. Mark Orsag/Dr. Amanda McKinney

Pathology—fever (1), myalgia (2), debilitation (3), overall rapid onset of symptomatology (4), bloody diarrhea/stool (5), esophageal wounds (6), frequent vomiting (7), blood in eyes (8), gangrene/damage to limbs (9), cases of loss or impairment of sight or hearing in survivors (10), extremely high case fatality rates (11).

Epidemiology—novel disease (12), "Ethiopian" origins (13), southeast to northeast (14) epidemiological dissemination pattern (15), cooler weather seasonality (16), infected corpses very infectious/deadly (17), abated but didn't disappear in summer(18), apparent human-to-human transmission (19),throughout Roman Empire in approximately to 2.5–4 years (20), affected the Roman military (21), possibly a universal pestilence hitting urban and rural areas (22), unprecedented and not recurring (23), pandemic temporal parameters approximately 249–262 CE (24), followed breaking of severe drought (25), seems to have smoldered between repeated massive urban outbreaks (26), hit some cities at least twice (27), generally indiscriminate (28), households and caregivers seem to have been particularly endangered (29), inspired great fear and caused massive depopulating flight from cities with large outbreaks (*ville morte* events) (30).

We will explore the ancient evidence for this list in an even greater level of detail in Part III. Many of the thirty points of pathology and epidemiology summarized here indeed very strongly fit with a retrospective diagnosis of a filovirus, or more broadly, with a VHF. Though, of course we will also analyze and consider potential alternatives. Harper's analysis, rooted in the aforementioned sources, places the plague's likely origin point beyond the Roman Empire's borders in non-Roman Africa ("Ethiopia") in 249 CE. It is important, as Harper notes, to not conflate ancient Roman (and inclusively Greco-Roman) use of the term "Ethiopia" with the modern nation state. Romans, in fact, used the term "Ethiopia" in multiple ways—in a broader ill-defined way to describe non-Romanized Africa. In ancient times, this included multiple civilizations:

Meroitic Nubia (with its capital city of Meroe located to the northeast of what is now the modern Sudanese capital city of Khartoum), the bordering Kingdom of Aksum to the East (with its Ethiopian Highlands capital city of Aksum) and, farther to the West, "Ghana" (another confusing ancient/modern conflation). More narrowly and precisely, it was used to indicate the geographical regions of landlocked Africa south of Roman Egypt that were the source regions of the north-flowing Nile River. From somewhere to the south and southeast, in Harper's conception, the plague seems "to have migrated north and west across the empire" [16, p. 137].

In our view, by far the most likely plague-origin candidate was actually the Kingdom of Aksum (located to the southeast of the Roman frontier town and port of Berenice) in the essentially Red Sea-defined Horn of Africa. Our addendum to Harper's hypothesis is based partly on general directionality and proximity, overall levels of trade and other contacts, and the crucial issue of effective control of the key Red Sea trade routes in the third century, as well as a number of other factors. More inland and Nile-defined Meroitic Nubia was also culturally and commercially linked with Roman Egypt, but Meroe lacked the crucial maritime commercial volume, speed, and ease of travel and access to the Red Sea, and through it, to the Roman Empire enjoyed in the mid-third century by Aksum. Considerable additional evidence supporting all these preliminary conclusions will be explored in much greater detail in our Parts II and III. Our conclusions drawn there will further bolster Harper's original temporal and epidemiological finding that, from this southeastern direction, the plague entered Roman Egypt and spread southeast to northwest and north throughout the empire.

In moving from a brief framing of the plague's likely point of origin to its potential pattern of dissemination, we must begin with Harper's original conception of the plague's temporal and epidemiological progression resting partially, if not fully, on the Dionysius–Eusebius account regarding the Empire's "second city" of Alexandria in Egypt as well as on some archeological finds made near the Egyptian city of Luxor. To say these are the only two evidentiary bases for his conclusions concerning "exotic" origins and an initial Egyptian epidemiological epicenter within the Roman Empire, as well as for his dating of the plague's Roman outbreak to 249 CE, however, puts it too strongly. The anonymous tenth-century Byzantine *Excerpta Salmasiana II* as well as several other contemporary and mutually supporting Greco-Roman and later Byzantine-era sources are also supportive in this regard. These later sources often recapitulate now lost third-century Roman sources in terms of noting the plague's "Ethiopian" origins and progression. Other such sources utilized by Harper include, but are not limited to, the third-century Neoplatonist philosopher Porphyry, the fourth-century Bishop St. Gregory of Nyssa, the tenth-century Byzantine historian Symeon the Logothete, the eleventh-century Byzantine historian George Kedrenos, and the twelfth-century Byzantine theologian and historian John Zonaras. Harper also deduced additional information concerning the Plague of Cyprian's peak fall-winter seasonality from some of these same sources—particularly that of Symeon. Harper's pioneering scholarly role in collating and linking these diverse sources simply cannot be overestimated; in many ways he "rediscovered" this "forgotten" ancient Roman pandemic [16, p. 136; 17].

The combination of all these accounts constitutes Harper's initial evidentiary support for the proffered southeast to northwest disease dissemination pattern. His account of the plague's "Ethiopian" origins had also been fairly conventional in terms of the, albeit limited, modern scholarship on the subject—as the collective verdict of the various ancient sources seemed largely unchallenged. Sabine Huebner, however, in her highly relevant 2021 article in the *Journal of Roman Archaeology*, interpreted these "Ethiopian pestilence" attributions as examples of an oft-recurring Greco-Roman xenophobic *topos* (traditional literary theme). Harper's contrasting interpretation of this and other evidence (particularly the Dionysius/Eusebius account in the *Ecclesiastical History*), however, leads him to what he believes to be an "incontestable" conclusion that the pestilence had spread southeast to northwest into Roman Egypt and was raging in the imperial metropolis of Alexandria by 249–250 CE. The Plague of Cyprian's characteristic symptoms were fever, myalgia, bloody diarrhea, vomiting, esophageal lesions, conjunctival bleeding from the eyes as well as damage to and putrefaction of the limbs [16, pp. 136–146; 17, 20, 21, p. 161] These descriptions are often drawn directly from Bishop Cyprian's eyewitness account, as the plague battered Carthage, in the early 250s CE:

> This trial, that now the bowels, relaxed into a constant flux, discharge the bodily strength; that a fire originated in the marrow ferments into wounds of the fauces; that the intestines are shaken with continual vomiting; that the eyes are on fire with the injected blood; that, in some cases, the feet or

> some part of the limbs are taken off by the contagion of diseased putrefaction; that arising from the maiming and loss of the body, either the gait is enfeebled, or the hearing is obstructed, or the sight darkened.... [6, p. 122 (14)]

Cyprian's and other ancient accounts also demonstrate vague perceptions of disease transmission through touch and emphasis on the eyes as a particular focal point of the plague. The terrifying pathologies also included damage to senses such as hearing and sight in survivors. Cyprian's account, for instance, mentions "disease of the eyes and the attack of the fevers." Cyprian also uses an allegorical biblical reference to Tobias, who "suffered the loss of his sight" in his "bodily affliction." Modern microbiologists and virologists have noted that members of the *filoviridae*, such as Marburg and Ebola, strongly present and can persist (even weeks after a putative recovery) "in the anterior chamber of the eye" and that "sensory changes" are noted in survivors. It is important to note, however, that many other viral infections—some well-known modern examples include severe acute respiratory syndrome (SARS) and the Zika virus can also medically present in a similar manner. Thus, while highly suggestive of a virus over alternatives such as a bacterial or a parasitic infection, our team concluded that this specific evidence is, in itself, not more narrowing to the members of the *filoviridae*. Other survivors, however, were described by Cyprian as suffering "feebleness" and "cruel harassment of the limbs." It thus cumulatively bolsters Harper's case that very similar pathologies are also noted in victims and among survivors of filovirus outbreaks in Sub-Saharan Africa today. In a similar vein, both Cyprian's and Dionysius/Eusebius' accounts emphasized the indiscriminate nature and extreme lethality of the disease. Handling the bodies of the dead was, according to the account of Dionysius, a highly "dangerous task." In Cyprian's words, "...many of our people are liberated from this world." For the former, the disease among the pagans was "more dreadful than any dread." Christians, whom the Dionysius/Eusebius account claims behaved in a more noble and supportive manner than their pagan neighbors, were nonetheless "continually following those who had gone before them [into death]" [6, pp. 120–122 (10–15); 9, pp. 135–139 (7.22.6–9) and (7.11.24); 10; 25].

Non-Christian sources echo, in regards to the question of lethality, this conclusion. Huebner, citing Zosimus' account of an outbreak during a military expedition that took place in 259 CE during the reign of the ill-fated Emperor Valerian, notes that more than half (unclear how many more than half) of the strong, fit, presumably well-fed (and with a strong standard of palliative care characteristic of the Roman military) soldiers nevertheless died of the disease. A similar verdict that death was the likely outcome was rendered by fourth-century writer Bishop Gregory of Nyssa. While an exact Plague of Cyprian case fatality rate (CFR) is, of course, undeterminable from such limited ancient evidence, the holistic evidentiary support for an extremely deadly Plague of Cyprian pathogen, that demonstrated strong analytical linkages to the basic pathology/epidemiology of VHFs (including filoviruses) is collectively irrefutable [17, p. 238; 21, p. 161; 37, p. 11 (1.20–21)].

According to Harper, major urban population centers in Africa, such as the great city of Alexandria in Roman Egypt, were hit hard and early. By likely 252 CE at the latest, the truly lethal pestilence was not only devastating Cyprian's Carthage but also had crossed the Mediterranean and was decimating the population of the city of Rome itself, as well as that of many other cities across the empire. Indeed, ancient sources such as the Codex Calendar of 354 CE, the later account of the eighth-century Monk Pseudo-Dionysius (who reported mass graves in Roman Syria) and the fragmentary surviving accounts of the relatively contemporary Greco-Roman historians Philostratus and Dexippus seem to pinpoint the years 251–253 CE as a period of very widespread and "massive" mortality within the Roman Empire from the Plague of Cyprian. Stressing the universality of the Plague of Cyprian, Harper describes the plague as having afflicted "cities and villages" and thus appears to side with more universalist ancient sources like Orosius ("... a great pestilence harassed the entire Roman Empire") rather than a source such as Zonaras who emphasized that the primarily urban plague "denuded many cities of their inhabitants." Such latter phenomena are often referenced in modern scientific epidemiological literature as *ville morte* events [17, pp. 234–237; 18, 22].

Through differential analysis, Harper also notes relevant silences in the ancient sources to eliminate or declare unlikely the prevailing explanatory paradigm of any and all types of smallpox (including the initially promising but relatively rare hemorrhagic variety). We fully concur with and have somewhat expanded upon his rejection of smallpox. Harper cites the lack of any description of the ever-present full body rash (which fits both the *variola major* and the more rare and lethal flat type of smallpox that disproportionately affected children), the ill fit of the pathology of permanent limb and sensory damage in survivors (absent in all three smallpox types) and convincing nearly contemporary ancient testimony (see below) that seems to indicate a "different disease" than smallpox. Additionally, no mention of a dusky erythema or evidence of pregnant women being particularly susceptible, both even more characteristic of the rare hemorrhagic smallpox than a rash, emerges from the ancient accounts [2, 16, pp. 109–141].

Harper also deftly undermines the prevailing Plague of Cyprian explanatory paradigm of smallpox by noting that such "fiery" ulcers, "appropriately called" "carbuncle[s]," are also explicitly mentioned by the very same Christian commentator Eusebius in his description of an early fourth-century plague in the *Ecclesiastical History*. Eusebius also explicitly identifies that pestilence as "another severe disease" marking it, in Harper's estimation and ours, as "a different" disease than the Plague of Cyprian that Eusebius had discussed (often through the medium of Dionysius letters) earlier in Book VII of the *Ecclesiastical History*. The only intervening reference to epidemic disease clearly presages the referenced fourth-century event described in that work's Book IX. The rash from the fourth-century pestilence began "spreading over the entire body." The disease "greatly endangered the lives of those who suffered from it"; yet "multitudes" survived it. The common smallpox *sequela* of post-infection scarring/lesions in survivors, while not explicitly described by Eusebius, can be very strongly inferred from the description provided. This disease, unlike the Plague of Cyprian, nearly perfectly fits the most common form of smallpox (*variola major*—or an ancestral lineage pathogen closely related to it). Contrastingly, not at any point in any ancient account regarding the Plague of Cyprian are "carbuncles" or, for that matter, the scars/lesions associated with survivors of smallpox mentioned. Harper's dismantling of the case for smallpox as the cause of the Plague of Cyprian is quite complete [2, 9, pp. 180–181 (9.8) p. 139 (7.22); 16, pp. 109–141].

Harper also notes the intense fear, something that seems to have resulted in unique partial/temporary or even permanent *ville morte* events, that the apparently novel and particularly deadly and terrifying Plague of Cyprian, incited in the empire's population. The plague's contemporarily noted exotic origins beyond the Empire's boundaries in "Ethiopia" also argues against smallpox which, by 249 CE, would have been a known and epidemiologically "endemic [if still ill-comprehended] pathogen" *within* the Roman Empire. Indeed, a later (sometime in the late 250s CE) metaphorical reference (… "what a plague spot of our thoughts…") from Cyprian, unconnected to his vivid description of the mid-third-century plague that bears his name, seems to confirm his awareness of such smallpox-like symptoms. This lends added weight to their absence in "On Mortality." The latter waves of the massive second-century Antonine Plague, in which smallpox is widely suspected to have been the pathogenic culprit, would have been under seventy years in the past by the mid-third century. Memory of that event was thus probably well preserved in both oral tradition and in written record [5, p. 163 (7)].

Before moving on to eliminate further pathogenic suspects, however, it is important to mention another important conclusion in this regard. The early fourth-century likely smallpox outbreak described by Eusebius "consumed entire households and families" and constituted a "severe threat," but seemingly more often than not, wasn't fatal. Historical CFRs from common (*variola major*) smallpox correspondingly seem to average around 30%. Some now-extinct pathogens in its ancestral lineage seem to have displayed somewhat, if not drastically different fatality rates. Contrasting descriptions contained in Eusebius' account thus seem to indicate that the Plague of Cyprian pathogen was far more terrifying than the disease (almost certainly *variola major* or a pathogen from its evolutionary lineage) implicated in the early fourth-century pestilence. In his descriptions of the Plague of Cyprian, Eusebius, using the eyewitness account of Dionysius, also speaks of a "multitude" but of "the dead and dying." This mid-third-century plague was more "dreadful than any dread and more intolerable than any other calamity… the only thing that prevails over every hope." Eusebius, in both instances, was a Christian describing suffering among hostile predominantly pagan communities, so the contexts were very similar in that sense. As for the Christian community during the fourth-century pandemic, Eusebius characterizes their convert-winning selfless assistance to their own and the pagan communities triumphantly. Such service included "burying the dead" and feeding those suffering from the effects of a contemporaneous famine. Additionally, Christians assisted the sick, recovering, and starving multitudes "who had no one to care for them." Once again, a different and more lethal reality seems to have prevailed among the Christian community during the Plague of Cyprian. "They [the Christian brethren] held fast to each other and visited the sick fearlessly… And they died with them most joyfully…And they took the bodies of the saints into their open hands…And after a little, they received like treatment themselves, for the survivors were continually following those who had gone before them [into death]" [9, pp. 139–181 (9.8 and 7.22)].

McNeill's alternative implication of the less deadly, highly aerosolized, and generally even more infectious disease measles, with its [unmentioned] characteristic blotchy rash and too recent genetic emergence (or inferentially its paramyxovirus relatives—the recently discovered [1994, 1999, 2012] Hendra, Nipah, and Sosuga viruses) is [are] also convincingly dismissed by Harper's or our differential analyses. Bubonic plague (no mention of swollen lymph nodes in Cyprian's or other accounts), typhus (including scrub typhus), dengue, yellow fever (mosquito-borne), etc., also don't fit the seasonality, epidemiology and pathology noted in List 2.1. Additionally, Cyprian's account makes no mention of anything like jaundice. Some tick-borne diseases are also counter-indicated by the seasonality (or simply the relentless nature of the Plague of Cyprian) and their being largely epidemiologically rural as opposed to urban. Harper rules out all the mosquito-borne flaviviruses due to the seasonality of the Plague of Cyprian. Once again too, a nearly relentless disease pattern characterized by large urban outbreaks would also eliminate this whole range of pathogens from consideration. Cholera lacks the hemorrhagic symptoms and had a biogeography restricted to India until the early nineteenth century [16, pp. 141–145].

Other disease candidates suffer a variety, in Harper's analysis, of "insuperable problems" (such as, once again, too recent molecular-phylogenetic emergence). Anthrax is also convincingly dismissed by Harper due to it being "not easily communicable between humans." A bit more tentatively, Harper identifies (while not entirely differentially eliminating it) a pandemic influenza, along the lines of the early twentieth-century Spanish flu, as a much less likely cause of the Plague of Cyprian than a filovirus due to, among other details, the lack of any reported respiratory symptoms; there is also the problem of the unlikelihood of such a disease causing the reported lingering sensory and limb damage in survivors. The apparent failure of the disease to recede more completely during the summer months can also, our team of authors believes, be added to Harper's list of negative differential evidence regarding pandemic influenza (this again seems to apply whether one doubts Harper's evidence regarding seasonality or not…). While our team of authors, like Harper, cannot *entirely* rule out pandemic influenza, we will develop additional evidence in Part III that we think even more strongly and broadly mitigates against pandemic influenza, or indeed any type of aerosolized pathogen, as the cause of the Plague of Cyprian. Due to more recent emergence and the prominence of respiratory pathologies, our team concluded that coronaviruses are also not likely suspects. Harper's historical version of differential medical and retrospective diagnosis, which we have further supplemented here, is one of the most quietly compelling parts of his entire analysis. He acknowledges the problems inherent in such attempts at retrospective diagnosis, and acknowledges that it, unsupported, cannot inspire "great confidence." Yet his analysis is, nonetheless, relentless, well-informed, effective in its interdisciplinary nature, extremely comprehensive and highly logical. One suspect pathogen after another is convincingly dismissed. We have built upon Harper's work here further by eliminating diseases such as the exclusively rodent-dropping borne [VHF] hantaviruses, due to their lack of human-to-human transmission, generally lower CFRs and interrelated lack of documented ability to produce large urban or temporally enduring outbreaks. We also eliminated unrelated bacterial q-fever due to generally exclusively rural epidemiology, lack of hemorrhagic symptoms, and (especially) lack of strong human-to-human transmission [16, pp. 141–145].

Factors relating to climate and ecology must also be considered when examining the holistic context of the Plague of Cyprian. One covered in great detail by Harper in *The Fate of Rome* and in an earlier 2016 article (*Journal of Roman Archeology*) is the fading of the Roman Climate Optimum (RCO) that had created a favorable situation for the empire's long and spectacular rise prior to 150 CE. Notable, in this context, was a megadrought and accompanying inconsistency of the annual June–September Nile flooding that apparently hit Egypt and Roman Africa (the empire's breadbaskets), the rest of eastern Africa, and the Middle East particularly hard in the 240s CE. Harper's sources, which range from early Christian writers (such as Cyprian and Dionysius/Eusebius) to Jewish rabbinical literature, to the papyrological records of Roman-Egyptian officialdom, are varied but, in Harper's analysis, consistent in their evidentiary verdicts. Between the early 240s and 248 CE, a "piercing drought" that produced a "present emergency" in Roman Africa and Egypt prevailed. So severe was the drought, in fact, that Harper (referencing the work of Dominic Rathbone of King's College in London) concludes that it was "the severest environmental crisis detectable at any point in the seven centuries of Roman Egypt." While Rathbone's analysis focuses more on horrendous "shortages" of wheat and grains, crops of all kinds seem to have "failed and wither[ed]." Food supply obligations to the Roman state, despite emergency requisition measures by the provincial government and the incentive of high prices from the state, weren't met. Overall, Dr. Rathbone also used papyrological evidence to document the drought and related famine in ways that strongly support Harper's conclusions concerning its severity. In a 2015 study, Rathbone concluded, in discussing the drought of the 240s CE, "This is the worst grain shortage attested in Roman Egypt." The tributaries of the Nile, even beyond the Empire's borders, apparently ran dry. This included the Blue Nile, which "gathers the runoff" from the Ethiopian highlands, and "carries it downstream to where it joins the regular flow of the White Nile [which also ran dry] near Khartoum." Granaries emptied and the cities of Roman Africa, the Middle East, and Egypt ran short of food. The Nile Valley, and even the riverbed itself, reportedly became, in the words of one ancient observer, as "parched as the desert" that surrounded it [16, pp. 129–138 and 332 (note 33); 31].

Indeed, there is also indirect support for the African and Middle Eastern drought's impact on the broader empire from a source not utilized in this context by Harper. In describing the situation regarding the overthrow and subsequent assassination of Emperor Gordian III in 244 CE by soldiers loyal to his successor Emperor Philip the Arab in 244 CE, the *Historia Augusta* blames "Philip's intrigue" for grain ships being turned away from ports on the northern side of the Mediterranean and "troops being moved to stations where they could not get provisions" [19, "The Lives of the Three Gordians" (29–30)]. While this passage is certainly suggestive of the drought's impact upon the entire Roman Empire, it is simultaneously marred by the potential bias and unreliability that often makes the *Historia Augusta* difficult for modern historians to use and interpret. The *Augustan History's* true authorship and sourcing are also uncertain. Scholarly debate over whether it is essentially a literary or historical source continues. Such issues, and the general paucity of applicable sources, generalized difficulties in precisely dating many ancient events and surviving sources, the confusing references in later texts to earlier sources now lost to time, convoluted calendar-conversion issues, and near constant use of factually imprecise religious allegory and metaphor in Jewish and early Christian-related sources also all complicate matters in dealing with Roman history in the third century. These are analytically limiting issues that we will attempt to supersede, in part, through our methodological multidisciplinary overlay of modern scientific evidence [33].

The fifth-century pagan historian Zosimus similarly relates the same mutinous event against Gordian III in his *New History*. Philip's Arabian origins seem to be a major factor in Zosimus' xenophobic hostility; he describes Arabia "as a nation in bad repute." But Zosimus' negativity was also almost certainly also due to Philip having adopted a tolerant, even supportive, attitude toward Christianity. Modern historical consensus accepts this impression of greater tolerance but simultaneously rejects the claims of Christian writers such as Eusebius and Orosius, author of *The Seven Books of History Against the Pagans*, that Philip or his son (and designated heir) actually became Christians. Zosimus praises Philip's strongly traditionally pagan successor Emperor Trajan Decius, who was by contrast, loathed by Christian commentators such as Orosius as a blood-stained "persecutor." A later passage related to the overthrow of Gordian III by Philip in the *Augustan History* is, however, more devoid of embellishment. It is straightforward on why the soldiers sided with Philip. It notes "that the anger of the soldiers against Gordian was due to hunger" [9, p. 121 (6.34); 19, "The Lives of the Three Gordians" (30); 29, pp. 314–315 (7.20); 37, p. 6 (1.13)]. Egypt and northern Africa were key imperial "granaries." Against this background of considerable analytical and evidentiary uncertainty, it is perhaps unsurprising that a work of the breadth and potentially paradigm-shifting nature of *The Fate of Rome* attracted criticism. As our own research and analysis is founded, at least initially, on Kyle Harper's pioneering concepts concerning the Plague of Cyprian (while also greatly expanding upon them), it is important for us to address the challenges by other prominent academics to Dr. Harper's scholarship.

The Fate of Rome's Detractors—Analyzing the Ancient Evidentiary Foundations

While the *Fate of Rome* garnered high praise from an impressive range of scholarly luminaries in the ancient history field and beyond including Dr. Ian Morris, Dr. Emma Dench, Dr. Peter Brown, Dr. J. R. McNeill, and Dr. Eric Cline (among many others), Harper's work also drew criticism from other highly distinguished scholars. Particularly vocal in this regard was a multinational/multi-institutional team of distinguished historians led by Dr. John Haldon (Princeton University), Dr. Hugh

Elton (Trent University), and Dr. Sabine Huebner (University of Basel). They published a three-part scholarly "response" to the book in 2018. Though they generally praised Harper's efforts in being the "first to attempt such a wide-ranging synthesis of the historical and scientific data pertaining to the Roman world," the authors strongly criticized what they alleged was evidentiary sloppiness regarding, particularly, the Plague of Justinian [12]. Regarding such issues as the Plague of Justinian, Harper's handling of the applicable climate science or the broader issue of the exact nature of the impact of climate change upon the rise and fall of the Roman Empire, much of their critique of Harper's scholarship is well beyond the narrower scope of our work. Haldon, Elton, Huebner, et al., however, also make several other criticisms that are more relevant in our context.

Some of the former include issues with Harper's overall approach—the *History Compass* authors charge insufficient multidisciplinarity, evidence being "distorted to fit the narrative," ignoring complex systems theory and insufficient attention to the challenges inherent in interdisciplinary consilience. They characterize Harper's analysis as more broadly colored by an "environmental determinist perspective" and overemphasis on the impacts and centrality of plagues (such as the Antonine Plague and the Plague of Cyprian) in Roman history. We will deal with some of the more relevant of these issues (interdisciplinarity and complex systems theory in particular) in our Part IV, but there are also some highly specific criticisms of Harper's work related to his treatment of the Plague of Cyprian that require immediate attention. The *Compass* authors believe that Harper's treatment of the issues of drought, famine, and the alleged irregularities of Nile flooding in the 240s CE, and more broadly in the third century, is flawed. They also criticize Harper's chronology and characterization of the outbreak pattern of the Plague of Cyprian as insufficient, incomplete, and unconvincingly sourced. They are particularly critical of his handling of certain source materials relating to the Dionysius/Eusebius continuum of evidence. Harper's critics believe this latter, in particular, to be a very important weakness in his case. They also allege analytically discrediting silences concerning the plague and the drought of the 240s in certain primary source materials, such as the third-century papyri from Roman Egypt and dispute Harper's handling of archeological evidence concerning the Plague of Cyprian found at the Funerary Complex of Harwa near Luxor. Huebner's 2021 article in the *Journal of Roman Archaeology* reinforced and greatly expanded this original critique of Harper's scholarship on the Plague of Cyprian in relation to these and other points [12, 21].

Huebner's 2021 "The 'Plague..." challenges Harper's Ethiopian-origin conclusion and the evidence that he presents for it very strongly; albeit with a central counter-argument that Harper had apparently anticipated in 2017:

> The pestilence came from Ethiopia and migrated north and west across the empire. So, the chronicles tell us, and we might suspect slavish emulation of the plague account in Thucydides, the model literary description of a plague, familiar to every educated Greek. But two telling clues corroborate the possibility that a microbial agent had invaded the empire from the southeast. [16, p. 137]

Huebner, however, challenges *both* of those subsequent evidentiary buttresses cited by Harper—the archaeological evidence from the Funerary Complex of Harwa site in Luxor and Harper's interpretation of the much-disputed Dionysius/Eusebius account. Huebner's argument is that the references in the various Byzantine accounts are indeed likely versions of this same xenophobic Greco-Roman "Ethiopia" plague *topos*, which had originated in the fifth-century BCE with the Greek historian Thucydides. She also, especially in the later 2021 article, collectively notes broader historical epidemic patterns that seem to cast doubt on the idea that the plague traveled from the African interior through modern Sudan up the Nile. She instead emphasizes the alternative potential role of the Red Sea ports [such as Myos Hormos and Berenice]—something that she believes Harper ignores. Our model of the plague will actually address a number of these points of Huebner's fully by the end of this work. In the shorter term, we can add two more points in defense of Harper's origin conclusions. This "Ethiopian" *topos of* Thucydides wasn't universally applied to all Roman disease outbreaks. It was missing, for example, during the Antonine Plague. If it was resorted to, in the case of the Plague of Cyprian, very plausibly, ancient authors applied it in response to contemporary perceptions (utilized later during the Byzantine period) in reaction to now lost (whether that of Philostratus or others) ancient source accounts which indeed implicated a southeastern pestilential origin. Another point made effectively by Harper, in a slightly different context, that we believe telling here is evidence of the Plague of

Cyprian's terrifying novelty. Citing the anonymous third-century Christian source *De Laude Martyrii*, Harper notes the described "...previously unknown" nature of the pestilence as evidence of the Plague of Cyprian's "exotic origins." The fact that the pestilence was different and unprecedented is also echoed, more generally, by such ancient sources as Zosimus, Orosius, and Pontius ("No one trembled at the remembrance of a similar event."). One can justifiably conclude that something unusual and perhaps quite sporadic (that was unprecedented and also apparently not repeated) had happened in an epidemiological sense. We thus find that Huebner's critiques *do not* undermine Harper's origin conclusions—at least at this point in our analysis. In her 2021 *Journal of Roman Archaeology* article, Huebner also advances a complex and fully developed alternative Plague of Cyprian origin theory of her own [16, pp. 98–140; 21; 37, pp. 8–9 (16–17); 29, pp. 315–316 (7.21–22); 30, p. 8 (9)]. We will examine it from our systems-synthesis perspective in Part III.

Huebner also invokes another Greco-Roman *topos* (...of disease ending in response to pagan sacrifices at the mid-to-late summer rise of the star Sirius) to cast doubt on the evidence in Symeon's account that Harper uses to deduce the Plague of Cyprian's seasonality. In this case though, the evidentiary value contained in Huebner's disputation seems limited. There are also more holistic systems-synthesis concerns involved. Huebner uses a sole (explicitly pagan) source remote in time (Diodorus Siculus in the first-century BCE) to dispute a later Christian account. The key difference between a disease ending (Diodorus) and abating (Symeon) also seems to be entirely glossed over by Huebner in this analytical context. There is also no mention by Huebner of any *topos* regarding disease arising in the Fall. Additionally, Huebner fails to respond to Harper's excellent point that the prevailing overall seasonality in the ancient Mediterranean tended to actually see an annual surge in gastrointestinal and other disorders in the late summer; so that Symeon's description of the Plague of Cyprian's seasonality actually reversed, rather than followed, traditional disease outbreak patterns. Furthermore, Plague of Cyprian outbreak peaks at Alexandria and Edessa in 260 CE, where seasonality can be clearly deduced, avoid the July–August period (further validating Harper [and Symeon]). We believe, as will be discussed later in Part I, that the same case can be made for an autumn beginning of the Carthaginian outbreak (which we date to 251–252 CE). Additionally, once again, if such a *topos* was applied, it might well have been used due to it matching or explaining the perceptions of contemporary observers [17, p. 244; 21, p. 163].

Other than Symeon's nuanced account, there are also no other surviving references to the plague's seasonality in the ancient sources. From our systems-synthesis perspective, such a dismissal also inferentially leaves us with the potential of an infectious disease with no seasonality—the "unusually continuous" pestilence described, for instance, by Orosius. This is something that recent twenty-first-century medical research tells us is effectively impossible: "Seasonality is a powerful and universal feature of infectious diseases… There are four main drivers of seasonality" that include environmental factors such as vector-borne disease (for example, those that infect humans through mosquitos) and the impacts of temperature and humidity. Human behaviors that interact with the epidemiological properties of a pathogen, quantified by a formula/infectivity-calculation labeled R0 (reproduction number), are another driver that we will discuss much more thoroughly in Part III. "Ecological factors such as algae play a role in cholera outbreaks." Seasonal biological rhythms may contribute to polio outbreaks. Some seasonal timelines, such as those for mosquito-borne disease, are [or would have been historically] quite obvious to those experiencing an epidemic, even if the causes were mysterious and unknown [23, 29, p. 316 (7.22)]. Epidemiological seasonality is thus inextricably linked, not only with human behavior, but also with the geographically prevailing conditions and cycles of planetary ecology. In sum, we thus believe Harper's use of the evidence in Symeon's account to be fully justified.

Harper's handling of the issue of the Funerary complex is based on the 2014 final report of the team of Italian archeologists (led by Dr. Francesco Tiradritti) who conducted the work at the site. Unless future expert on-site archaeological work is conducted at the site that effectively undermines the conclusions of Tiradritti's team, Harper also seems fully justified in using Tiradritti's conclusions as solid evidence supporting Harper's own. On other points, we found Huebner's (as well as those of the other *History Compass* authors) competing characterizations largely persuasive or differed from them only in highly nuanced ways. Her complex and knowledgeable analysis of the applicable numismatic evidence, which she claims Harper over-interpreted, can perhaps be simply characterized by her eminently defensible verdict of "possible but hardly definite." We will, however, revisit this issue somewhat in Part III as we do see the numismatic evidence as a limited but still analytically viable puzzle piece when seen in context. Huebner also disputes Harper's handling of certain archaeological (a mass grave from the catacomb of Saints Marcellinus and Peter in Rome) evidence. Here, the issues become very murky indeed as new archaeological interpretations of the materials at the site have been published since Harper's relevant scholarly works, identifying the site as a potential mass grave for Plague of Cyprian victims, appeared. In the intervening years, archaeological work at the site, in some ways, has stalled and in other ways progressed. On balance, we believe that Harper's overall case regarding the Saints Marcellinus and Peter Catacomb has actually gotten somewhat analytically stronger over the last several years. This, deduced from holistically analyzing the evidence presented by Huebner in 2021; though conclusions, on the whole, remain (exactly as Huebner argues) less than fully definitive. Furthermore, any differences of opinion on this narrow and complicated evidentiary matter between Huebner and Harper change the holistic analytical picture regarding the Plague of Cyprian not at all. Huebner concurs with both of Harper's conclusions that the Plague of Cyprian was indeed a pandemic, and that it heavily impacted the city of Rome itself on multiple occasions [16, pp. 136–138 and 332 (note 38); 21, pp. 168–169].

We believe Huebner's use of the relevant Egyptian papyrological evidence, however, to be highly erudite and revelatory. Her 2021 article in this regard proved immensely useful to us in ways that we will discuss far more thoroughly in Part III. One other interrelated point from the original *History Compass* authors also proved wholly persuasive. Some of the data on Nile flooding over the years referenced or utilized by Dr. Harper, at times in conjunction with his mentor at Harvard Dr. Michael McCormick, whether derived from secondary sources or generated independently, seems (exactly as Haldon, Elton, Huebner et al. contend) unconvincing in terms of matching evidence with analysis. It was also, for our purposes at least, too imprecise or confusing to cite or utilize in any way in our present work. If the Nile did indeed abandon the Romans "spectacularly," in the third century as Harper and McCormick have contended, this conclusion has yet, in our view, to be proven [We were, for example, unable to conclude that McCormick's included Nile flooding charts adequately supported the conclusions drawn from them… 12, 16, p. 134; 21, pp. 165–168; 26]. The narrower, and in our context more important, issue of the drought of the 240s CE, however, is a different and more nuanced issue. While more definitive scientific data on the annual flooding patterns of the Nile during the third century may indeed not be available until at least 2024 (and the planned completion of a project under the leadership of the same Dr. Sabine Huebner of the Institute of Ancient History at the University of Basel), close reading of multiple ancient sources and a synthesis with recently developed modern scientific evidence related to infectious disease does ultimately provide some compelling clues. These ultimately strengthen Rathbone's and Harper's overall case, albeit with certain caveats, regarding the drought of the 240s CE. The mutually reinforcing support provided for Harper's conclusions regarding famine, drought, and pestilence in the 240s CE is, in sum, at least very highly suggestive. While some of this system's synthesis evidence regarding drought and famine is burdened by certain intrinsic uncertainties and lurking potential circularities (particularly if viewed less holistically), its overarching and cumulative impact ultimately creates consilience and a powerful meld. Much of this same evidence is also closely intertwined with Harper's chronology of the Plague of Cyprian in Alexandria (with its chief evidentiary basis resting partially, if not wholly, on the controversial account of Dionysius/Eusebius). We will thus handle these two interrelated and, in context, crucial issues together.

One good starting point is with a relevant historical event that we are surer of in a temporal sense; the so-called "Decian Persecution" that began in Egypt (and indeed across the Roman Empire) in late 249 CE. After Emperor Philip the Arab was overthrown in 249 CE by Gaius Decius, the latter instituted an empire-wide course change. A firm traditionalist, Emperor Decius attempted to stabilize the increasingly troubled empire by restoring pagan cultural and religious unity. He adopted the name Trajanus, after the great early second-century emperor. With a dose of public patriotism and a demonstration of pious loyalty from the population, he seems to have believed the "peace of the gods" would return. He issued a decree in the fall of 249 CE legally mandating a documented public show of both political loyalty to the emperor and adherence to traditional pagan religious practices. Legal certificates from the campaign (from as early as January 250 CE) have survived. An example appears below:

> To the Commissioners of Sacrifice of the Village of Alexander's Island: from Aurelius Diogenes, the son of Satabus, of the Village of Alexander's Island, aged 72 years: --scar on his right eyebrow.
>
> I have always sacrificed regularly to the gods, and now, in your presence, in accordance with the edict, I have done sacrifice, and poured the drink

offering, and tasted of the Sacrifices, and I request you to certify the same, Farewell.

Handed in by me, Aurelius Diogenes

I certify that I saw him sacrificing...

Done in the first year of the Emperor, Caesar Gaius Messius Quintus Trajanus Decius, Pius, Felix, Augustus, [June 25 250 CE]. [3]

All imperial subjects were thus required to sacrifice to the traditional pagan Roman pantheon and for the emperor's health as part of a "unifying policy." While certain pragmatic and traditional accommodations were made for a well-established minority like the Jews, Roman governance seems not to have adopted such a lenient attitude toward Christians, whose constant proselytizing seems to have irritated and unnerved Roman authorities. For Christians, Decius' royal decree was an "important turning point"—a shocking first distinctive event signaling a descent into a pattern of persecution of Christians by Roman authority. Modern scholarship trends, however, tend to downplay the notion of anti-Christian "persecutions" under Decius. It is argued that such prescriptions were nearly universally applied and weren't actually as much of a break with previous practices as Christian sources claimed. Christians weren't, it is also argued, specifically targeted, and in a narrow rather legalistic sense that may have been true. The sacrifice was also for the emperor and not to him (as a deity); Decius wasn't Caligula. Such niceties were, however, certainly lost on the third-century empire's Christians. Christian teaching strongly forbade such practices, "He that sacrifices unto any gods, save unto the Lord only, shall be destroyed." Christians could resist, choose martyrdom, or face imprisonment, torture, judicial or extra-judicial execution (like Pope Fabian in Rome); or flee and face confiscation of their property (Cyprian chose this option). They could also attempt to get out of trouble by bribing a corrupt or sympathetic Roman local official to issue a fake certificate. Many other terrified Christians across the empire submitted and sacrificed–in Cyprian's words, "...conquered before the battle, prostrated before the attack. They ran to the marketplace of their own accord...How many were put off by the magistrates at that time when evening was coming on...." Here, Cyprian seems to be referring to the situation prevailing in Carthage at the height of Decius' campaign (around April 250 CE). Later debate within the Church about what exactly to do about these "lapsed" Christians continued long after Decius' campaign slowed down and then fully lost momentum toward the end of 250 CE. The emperor's subsequent death in battle against the Goths during the following year caused the legal decree that had created the campaign to lapse altogether. This event was noted as a "recent matter," with some grim satisfaction, by Cyprian (news of battles, plague outbreaks, and other events would have spread slowly over weeks and months around the empire) in his work "An Address to Demetrianus." Very importantly in our context, this almost certainly dates that literary work to the latter half of 251 CE [4, p. 101 (17); 7, p. 54 (7–8); 32].

As noted, Eusebius' *Ecclesiastical History* often draws upon, in the relevant sections, the contemporary accounts of Bishop Dionysius of Alexandria. Eusebius, in fact, directly notes the considerable archive of a variety of types of letters written by Dionysius that were available to the former in the fourth century. "And he [Dionysius] communicated with many others by letters, which he has left behind him as a benefit, in various ways, to those [a clear reference to Eusebius himself] who now diligently study his writings." At times, however, the "voices" of the third- and fourth-century writers can become difficult to distinguish in the *Ecclesiastical History*. It is Book VI which gives a brief account of Decius' alleged motive for initiating the campaign—"...his hatred of [his more pro-Christian predecessor] Philip." Dionysius' flight, four days after its initiation, from the "Decian Persecution" and the sufferings he endured are also discussed. The sufferings of Christians in the city of Alexandria immediately before and during the Decian Persecution—"The Martyrs in Alexandria" are described by Dionysius in a letter to Fabius, Bishop of Antioch. Our analysis will return to consider this crucial latter section presently. It is important, at this point, to further broach the complex and often jumbled, not clearly or consistently chronologically ordered and confusing nature of the *Ecclesiastical History* as a historical source. The highly relevant, in context, Books VI and VII include, often widely separated, single or multiple chapters and chapter groups addressing the same subject or event. The Decian Persecutions, for example, are the subject of Chapters 39–42 of Book VI and Chapter 1 of Book VII. The reign of the emperor Gallienus (r. 253–268) is the subject of Book VII's widely separated Chapters 13 and 23. In each case, other material was inserted by Eusebius between these interconnected sections [9, pp. 107–149 (6–7)].

On the other hand, Cyprian's accounts and language, though often allegorical, are also more straightforward and highly suggestive that the plague, while well before its peak "build" in the city in early 252 CE, was already raging in the empire during the extended period of the so-called Decian Persecution of 249–250 CE and its immediate aftermath. This evidence, in our opinion, can (independently of Eusebius' account) be used to, at least tentatively, bolster Harper's heavily contested timeline for Alexandria's plague outbreak as well as his conclusions regarding the plague's overall southeast to northwest spread. Cyprian links the sacrificial offerings to "plagues." The "plague-bearing idol meats" produce "fatal contact" and a "deadly blow that is planted in the deep and secret entrails." The words "health" and "medicine" are also used. Yet these are rather general, references that don't seem specific or immediate at least physically (or perhaps contemporaneous temporally) to Cyprian's Carthage in late 249 and 250 CE. It is plausible, if not proven, that Cyprian was referencing distant but recently past or ongoing epidemics in the major Roman African metropolis of Alexandria or elsewhere in the empire. There are also, however, probably later and more immediate references that seem to indicate the arrival of the plague in Carthage, such as "...these plagues that occur." Additional surprisingly specific references, albeit buried within a more general and allegorical passage, seem to document that by the time Cyprian wrote these words, the Plague of Cyprian was past already its epidemiological beginnings and had begun to build in the Roman African city of Carthage. It had not yet, however, reached the near-apocalyptic levels described in "On Mortality." "You reproach plague and disease, while by plague itself and disease, the crimes of individuals are either detected or increased, while mercy is not manifested to the weak, and avarice and rapine are waiting open-mouthed for the dead... shunning the deaths of the dying, and craving the spoils of the dead... he who enters so eagerly upon the estate of the dying, probably desired the sick man to perish" [4, pp. 96–98 (7–8 and 10); 7, pp. 55–58 (7–8)].

The latter references were, once again, written in "An Address to Demetrianus" probably around October–November 251 CE. It is notable that this fits very well with the overall seasonality described by Symeon. But how long was this before Cyprian authored "On Mortality" (the work that seems to have coincided with the plague's peak in Carthage)? Huebner dates "On Mortality" to 253 CE. McNeill had earlier dated it to 251 CE. Harper's scholarship dates it to 252 CE. Other scholars suggest a broader range of 252–254 CE. Exact timing, due to Cyprian's use of past tense and a series of cascading temporal uncertainties hardly uncommon when considering the timelines of ancient events, create a less than crystal clear picture. Due to the consilience provided by evidence of other Plague of Cyprian peak outbreak timelines from Alexandria and Edessa, however, and through broader overall interdisciplinary analytical consilience, we estimate roughly that "On Mortality" was likely written in the late winter/early spring (likely no later than February or early to mid-March) of 252 CE (before the vernal equinox). There is, after all, no mention of Easter-related activities, as in Dionysius/Eusebius' account for Alexandria, so an analytically proffered date of later in the spring (April or May) might well be deemed too late. The evidence of Symeon's seasonality seems to largely rule out a later summer peak. This, of course, is all slightly speculative and contingent, but a number of analytically adjacent disparate pieces do seem, collectively, to denote the outlines of a missing piece of the Plague of Cyprian puzzle [1, p. xiii; 17, p. 229; 21, p. 160; 27, p. 108].

There are yet further reasons why this particular dating makes the most sense to us. The reign of Emperors Gallus and his son Volusianus, in Orosius' words, was "famous for this plague alone." In contemporary and later Roman perceptions, the pestilence characterized and dominated their two-year (summer 251–summer 253 CE) tenure. A number of other ancient sources (such as the fourth-century pagan writer Eutropius) make nearly identical characterizations. Thus, a temporal centering of a months-long major outbreak peak such as that in Carthage to 252 CE, as opposed to the co-Emperors' two-year reign's beginnings in 251 CE (McNeill) or end in 253 CE (Huebner), as Gallus and Volusianus were about to be overthrown by Aemilianus, is fundamentally far more logical. Even more tellingly, it will be argued in Part III that our epidemiological analysis also strongly indicates that the first wave of the Plague of Cyprian's outbreaks in Rome and Carthage *almost certainly overlapped*. This considerably strengthens the conclusion above; events in the city of Rome itself would have been widely seen as greatly significant and defining. Collectively, we believe this evidence initially suggests overlapping Plague of Cyprian outbreak dates of around September or October 251 CE with epidemiological peaks coming in early 252 CE for both Rome and Carthage. The probable epidemiological "burn and build" timeline based on this analysis (other factors regarding Mediterranean sea lanes and maritime travel times that will be discussed in Part III figure into this calculation as well) for Carthage, a significant imperial city of around 102,000 people in the third century, we calculate as being around 6–7 months. That, if even close to correct, also seems at least comparatively too slow for the epidemiological progression of more easily disseminated aerosolized pathogens in a densely populated ancient urban epidemiological setting lacking effective disease mitigation measures. Third-century Carthage had easily more than twice the population density of modern New York City—though obviously not the interconnectedness and ease of movement that is a product of modern technology [13, p. 319; 17, p. 240; 24; 29, p. 315 (7.21); 34].

The toll in Carthage was grim indeed; Cyprian noted that "…many of our people die in this mortality." Widespread flight from the city (a partial *ville morte* event) also occurred. According to the account of Cyprian's associate Pontius, those that remained behind fell ill and died, indiscriminately, at what seems to have been a terrifyingly high CFR. Cyprian himself noted, "This mortality, as it is a plague to Jews and Gentiles, the enemies of Christ, so it is a departure to salvation to God's servants." The plague thus reversed the urban Roman Empire's annually predictable late summer urban peak mortality disproportionately affecting children [6, pp. 119–123 (8–15); 15, pp. 15–34; 30, p. 8 (9)].

In a 2016 article, Harper also analyzed another contemporary anonymous Roman African Christian source—the work entitled *De Laude Martyrii* ("In Praise of the Martyrs") that seems to refer to the entire span of time directly leading up to the deadly 252 CE urban epidemiological peak when "the carcasses of many" lay in the streets of Carthage—the leading city in non-Egyptian Roman Africa:

> Or do we not see the rites of death every day? Are we not witnessing strange forms of dying? Do we not behold disasters from some previously unknown kind of plague brought from prolonged and furious diseases? And the massacre of wasted cities.... [8, 14]

Such a commentary evidences widespread *specifically epidemiologically urban destruction* from the Plague of Cyprian in the Roman Empire. Our List 2.1's epidemiological characterization of the plague as devastating but indiscriminate, exotic and terrifying, unique/not recurring, extremely lethal, slowly burning but building toward a terrifying peak and resulting in a very high urban death toll all find strong support in Cyprian's account and others cited by Harper. Pontius the Deacon refers to the urban outbreak in Carthage killing "numberless people every one from his own house." Harper cites "Apollo the Healer" and other religious imagery from Emperor Trebonianus Gallus' (Decius' successor) coinage from the central Roman mint in 251 CE (both the *antoninianus* and *sestertius*), and comments from the third-century pagan commentator Porphyry that imply a linkage between the Plague of Cyprian and the Decian Persecutions. It is quite possible, though not necessarily strongly evidenced, that the Decian Persecutions were, in fact, a *response* to the initial outbreak of the plague in 249 CE in Alexandria and elsewhere. This new "mortality of the age" was perhaps linked, in the mind of Roman traditionalists, to Christian refusal to "respect the gods." The numismatic imagery from 251 CE is indeed, exactly as Huebner argues, hardly definitive. But it is, in our view, plausibly reflective of an already Empire-wide pestilence by then. On the *antoninianus*, Gallus' bust appears on the front with Apollo the Healer on the reverse. On the *sestertius*, Gallus' bust on the front is complemented, on the reverse, by an image of the feminine health and well-being deity Salus. Additionally, the *sestertius* image religious/miraculous invocation notably lacks the image of the god of medicine Asclepios (which had appeared on similar coins in the previous decades). This may indicate a perceived failure of the urban Roman Hippocratic-Galenic medical system to deal with the spreading Plague of Cyprian—an issue that will also be dealt with in much greater detail in Part III. Another passage from Cyprian deals with the discomfort among Christians that the indiscriminate pestilence was every bit as lethal and devastating to them in 252 CE as it was to the Pagan and Jewish communities in Carthage. "But nevertheless, it disturbs some that the power of this disease attacks our people equally with the heathens…." [We note that our AHA co-panelist Florida State University doctoral candidate Baylee Staufenbiel provided some useful insights for this section during her February 2022 presentation, 6, p. 119 (8); 11; 13, p. 319; 14; 21, pp. 163–165; 30, p. 8 (9)].

Such words also seem to prefigure later Christian attempts, such as that of Orosius, to portray the plague as divine vengeance for the Decian Persecutions. Orosius condemned the murder of Philip and his son by Decius, whom Orosius denounces as "the instigator and repressor." Orosius characterizes the Plague of Cyprian's massive spread under Emperor Trebonianus Gallus and his son and co-Emperor Volusianus as "vengeance for the violation of the Christian name…" [29, p. 315 (7.21)]. Such a desire may have led Christian commentators in the direction of strategic silences or possibly more active "reassignment" of the plague's outbreak to the reign of Decius or, after his death in battle in

251 CE, to that of the following, also anti-Christian, Emperor Trebonianus Gallus. In some sense, it just wouldn't do for Christian writers to have the plague break out under the more sympathetic Philip the Arab in 249 CE, rather than under the oppressor Decius.

Cyprian's accounts of the period are also replete with references to drought and famine. "If the rain comes down with unusual scarcity, and the earth falls into neglect with dusty corruption... the fountains are now less plentiful to you...You complain of barrenness and famine" [4, pp. 96–98 (7–10)]. Cyprian seems to be referring to events affecting Carthage in the 240s and/or early 250s CE. Some strong additional corroborating evidence on these issues comes from ancient Jewish rabbinical literature and concerns a leading Rabbi and Talmudic scholar in the city of Sepphoris (in modern Israel) named Hanina bar Hama (who apparently experienced and survived both the regional drought and the subsequent pandemic before dying of old age). Rabbi Hanina is discussed by Harper in *The Fate of Rome*. Harper notes that the drought of the 240s CE is described in Jewish sources as withering, and it appears that ritual fasting failed to alleviate it. Yet, as noted by Harper, Hanina bar Hama (who probably died sometime in the early 250s CE) lived to see the rains more fully return. He also, and this evidence will be explored more thoroughly in Part III, apparently survived at least the initial wave of the Plague of Cyprian that devastated the Judaean city Sepphoris. There don't seem to be any other temporally close plague candidates, other than the Plague of Cyprian, prior to the latter's potential 249–253 CE outbreak continuum. Cyprian's biographer Pontius commented, "No one trembled at the remembrance of a similar event." As noted in points #12 and 23 of List 2.1, the Plague of Cyprian was also apparently a ***novel and unique*** disease event [16, pp. 131–132; 28, 30, p. 8 (9)]. Due to the usual problems in terms of precisely dating many ancient events, none of this is completely certain. Yet, the sum total of all of this evidence cumulatively fits with Harper's analysis of a unique disease event with an exotic "Ethiopian" origin, a southeast to north/northwest spread, proximity of the plague outbreak to drought, etc. His timeline, pathology and epidemiology, as more expansively rendered in our List 2.1, are also strongly supported.

Roman African and Middle Eastern population centers were also characteristically connected by sea lanes and imperial roads (though overland civilian travel times between Alexandria and Carthage, for instance, would have been around three months). Alexandria and other important regional centers were linked via waterborne trade routes (whether sea or riverine). There were also significant, both waterborne and overland, trade routes connected to, or passing through territory or waters (the Red Sea in particular) increasingly controlled in the third century by the Kingdom of Aksum. Seaborne trade across the Mediterranean also connected Africa to the rest of the Roman Empire. Waterborne trade was more rapid and cost efficient and thus predominated. An extremely important connecting hub in this ancient movement and commercial web was the city of Alexandria—the precise point where Haldon, Elton, Huebner et al., most strongly attack Harper's Plague of Cyprian outbreak chronology relating to the key ancient account of the *Ecclesiastical History*. We will review the evidence relating to that issue presently. A further examination of the contribution of relevant Jewish rabbinical literature to our understanding of mid-third-century drought and plague, and a potentially crucial connection between them, however, is necessary beforehand. One source from that tradition, not cited by Harper, also considerably buttresses our developing and interconnected case in multiple senses and provides an important additional puzzle piece.

That source is the *Tractate Taanit's* Chapter 3. The *Tractate Taanit* is composed of ancient Jewish definitions, rules, and regulations concerning droughts (and their breaking) and other threats in response to which ritual fasting might take place or alarms sound. It is, like much of the rest of rabbinical literature, composed of *Mishnah*, Jewish traditions, laws, and regulations that (in total) cover nearly all aspects of life and *Gemara*. The latter consists of interpretation, elaboration, and commentary on the *Mishnah* by the prominent rabbis of various eras, based on their experiences and beliefs. The apparently contemporaneous mentions of, and commentaries from, both Rabbi Hanina bar Hama and his famous pupil Rabbi Johanan bar Nappaha (the original compiler of the Jerusalem Talmud) in one section of the *Tractate Taanit's* Chapter 3 clearly reference a quite precise period of overlap (the 240s to the early 250s CE) in the fame of both the elderly teacher and the rising pupil. Rabbi Johanan would also go on to survive the Plague of Cyprian, perhaps employing practices learned from his religious mentor that will be detailed in Part III. While the organization of the *Tractate Taanit's* Chapter 3 is not strictly linear in a chronological sense, the commentaries are generally informally sorted according to a temporal order of religious/scholarly prominence. Rabbi Hanina bar Hama (death c. 250 CE), for instance, is mentioned in the same sections as both his mentor Judah I (death c. 217 CE) and his pupil Johanan (death c. 279 CE). The commentary from Hanina tends to follow the commentary of Judah, and the commentary from Johanan tends to follow the commentary of Hanina. The commentaries of Judah and Johanan, however, are generally not co-mingled. The commentaries of the later Babylonian Rabbi Rav Huna, who died around 297 CE and whose fame overlapped with Johanan's, are commingled with his but not with Hanina bar Hama's. The *Gemara* of the *Tractate Taanit* and the broader Jerusalem Talmud are thus mutually supporting in regard to this issue of temporal placement. A passage from the latter references this very period, late in Hanina bar Hama's life, after his star pupil Johanan had relocated away from the city of Sepphoris to the city of Tiberias. Upon seeing a crowd of people on their way to hear Rabbi Johanan bar Nappaha speak in Tiberias, Hanina bar Hama commented, "Praised be the Lord for permitting me to see the fruits of my labors before I die" [35–36 (*Horayot*)].

Another interesting feature of this section of the *Tractate Taanit* is its specific and straightforward quality. Precise rules and regulations are based on lived experience. A pestilence, requiring the sounding of alarms, is exactly defined... "in a city capable of furnishing 500 able-bodied men" when "3 [or more] persons die in three consecutive days." Drought is a "general plague causing dearth" when forty or more days pass between rains. Rain types too are cataloged—light, mild, sufficient, felicitous, liberal, gentle fructifying, heavy, adequate or inadequate, doing good for the sprouts or not. Progressions of the withering of crops are also discussed. Other threats requiring the sounding of alarms are defined as types of agricultural blight, infestations of locusts or crickets, or attacks of wild beasts or armed men [35]. One story that we passed over in this evidentiary context concerned the semi-mythical first-century BCE figure Honi the Circle Drawer, who once supposedly slept for 70 years. This source thus reads a bit like a Jewish version of *The Twelve Tables* (often considered a foundational document [actually a set of bronze tablets] of Roman law) with the benefit of additional relatively contemporaneous historically informative annotation. Allegorical and metaphorical language and religious mystical or mythical elements aren't altogether absent but also aren't particularly problematic in utilizing the *Tractate Taanit's* Chapter 3. One can exclude such sections in an evidentiary sense without losing analytically informative value. The strong overall impression that emerges from the text is that drought and food shortages were a recurring problem in the 240s CE in Judaea. In this sense, the *Tractate Taanit* strongly confirms the broader conclusions of Rathbone and Harper. The conditions that apparently regionally afflicted Egypt and northern Africa during the same period find a strong echo here; drought was clearly a regional phenomenon. Much of Chapter 3 of the *Tractate Taanit* is composed of discussion by the two famous religious scholars regarding situations from this period of varying degrees of combinations of food shortages and high prices. The dialogue is referenced with the aim of defining a situation of mere "want" (food shortages leading to higher but still manageable grain prices) as opposed to a true state of famine (in only the latter case should the alarms be sounded). Food could also be altogether absent or procurable only (and therefore far more expensive) via overland (literally "by asses") routes; either of those situations was deemed to be a true state of famine [35].

Situations where local food shortages existed, but grain was procurable from other parts of the Roman Empire via ships, were treated as more complex definitionally. If the Jewish population generally had the financial resources to acquire the food, then Rabbi Hanina bar Hama opined that such a situation did not require the sounding of alarms since a mere state of "want"—as opposed to a true famine existed: Hanina stated that "Even if the price of grain has reached the sum of one Sela for one Saah, but at that price it is obtainable, this would merely constitute a case of want." Johanan bar Nappaha is, however, subsequently quoted as explaining that, apparently at a particular but unspecified stage in the megadrought of the 240s CE, "I recollect it happened one time in Tiberias that although four Saah of grain were to be had for one Sela, still many people were starving to death, because they did not have the necessary coin with which to purchase their grain." In Johanan's view, such a situation indeed reflected a state of famine calling for the consequent sounding of the alarms. The experience of the breaking of drought is annually linked, at the latest, in the *Tractate Taanit* to the religious holidays of Passover (late March–early April) and the Feast of Tabernacles (late September–early October). The latter despite the fact that rain during the Feast of Tabernacles itself was regarded as unpropitious [35]. In a broader regional, Egyptian and Ethiopian geographical context, this latter timing overlapped with the traditional end of the Nile flooding (July–September) season. Such markers were linked to rules calling for the sounding of alarms if the breaking of droughts had not occurred by these holidays. These represented annual temporal/seasonal/ecological/climatic endpoints—with the drought-breaking rain events expected (or at least hoped for) sometime in the windows of time closely preceding them.

The text also discusses various types of agriculturally sufficient or partially sufficient rains, but that apparently is not what was necessarily experienced during this time period in these annual precipitation windows, broadly defined here as February–April and August–October. Drought, probably multiple times, in the era seems to have broken suddenly, violently, and less than ideally in an agricultural sense. Hoping for, praying for, or fasting for a moderate and sufficient rain that would nourish the sprouts and fill the wells and cisterns, however, would not necessarily bring it. "...How can that take place after what we have learned in a *Baraita* [Jewish Oral Law], namely: that sufficient rain descended to fill the wells, cisterns and caves, which was, however, inadequate for the trees and sprouts? That *Baraita* refers to a rain which fell in torrents and filled up the wells, etc. but did no good to the sprouts." Such events, however sudden and dramatic, wouldn't have, at least in the immediate term, strongly revived agricultural production. The text following the lesson/application of the *Baraita* is an accompanying mention of regulations concerning the threat of fatal infectious disease outbreaks. These are described as colds, whooping cough, or croup. Interestingly, though he isn't mentioned in that precise section of the text, such terminology reflects the, for the time period, unorthodox, medical belief system of Rabbi Hanina bar Hama. Hanina, in addition to being a noted Rabbi, was also an oil merchant and a respected physician. He seems to have generally rejected prevailing humour-based Greco-Roman medical theories and instead operated on the basis of a belief that almost all fatal infectious disease was caused by respiratory ailments such as colds, croup, and whooping cough. In that context, he stressed the use of bathing and special oils, which he also sold, that had been used on him (as a child) by his mother [35–36 (*Shab*)].

While this last, frustratingly inconclusive, bit of evidence is indicative that once again we are not dealing with a definitive source in the *Tractate Taanit*, the overall level of additional support for Harper's and Rathbone's characterization of the drought and resulting food shortages is very strong. The evidentiary interdisciplinary consilience issues raised by the *Tractate Taanit* are, however, thought-provoking as well. How could un-supplemented modern scientific analyses (alone) of regional precipitation and flooding levels over 1,700 years later even begin, no matter how technologically advanced and ingeniously derived, to reveal the subtleties and timing of fructifying or non-fructifying, moderate or heavy, good for the sprouts or not good for the sprouts, rains? The power of this highly detailed source is also multiplied synergistically, in our context, when one examines it alongside one of the most crucial (and clearly the most controversial) sources for Harper's overall analysis and timeline regarding the outbreak of the Plague of Cyprian in Alexandria. This source, of course, is Eusebius (and his oft-disputed recapitulation of the eyewitness account of Bishop Dionysius of Alexandria). This relevant section of Eusebius' *Ecclesiastical History*, Book VI Chapter 35 to Book VII Chapter 23 is, as noted earlier, somewhat chronologically confusing, oddly repetitive, and jumbled. Yet, it is also highly detailed and contains some very clear temporal markers. Using these, we contend that it is possible to establish a strongly convincing timeline for the outbreak of the Plague of Cyprian in Alexandria through this source as a key part of our broader evidentiary/analytical mosaic. This is, however, a challenging and varyingly interpreted source that has sparked vigorous and long-running scholarly disputes. In the concept of our broader analysis and modeling of the Plague of Cyprian, this source is, however, a key puzzle piece that simultaneously constitutes a puzzling conundrum in and of itself! Thus, we will analyze it in great detail. The issues here are difficult, but we strongly believe that it is possible to painstakingly "decode" the relevant sections of the *Ecclesiastical History* when they are examined more holistically. We thus begin our next chapter with our detailed analysis, followed by a summary of Huebner's (quite contrasting) analysis and conclusions. We will conclude this narrowly focused section of the following chapter concerning the Dionysius/Eusebius account with our evidentiary rebuttal to Huebner's conclusions.

References

1 Campbell, Philip and Grant, Ryan, 2013 *The Complete Works of Saint Cyprian of Carthage*, xiii. *The Complete Works of Saint Cyprian of Carthage* Translation Editing: Nicene Fathers [1885] Philip Campbell, Ed. Introduction by Ryan Grant, (Evolution Publishing), pp. v–xvi.
2 CDC, 2020 "Smallpox, Clinical Disease," *Centers for Disease Control*. https://www.cdc.gov. Accessed on 5/11/2021.
3 "Certificate of Having Sacrificed to the Pagan Gods, Egypt, 26 June AD 250," in Lewis, Jon, ed., 2003 *The Mammoth Book of Eyewitness Rome* (Carroll and Graff), p. 378.

4 Cyprian, 2013 "An Address to Demetrianus," *The Complete Works of Saint Cyprian of Carthage* Translation Editing: Nicene Fathers [1885] Philip Campbell, Ed. Introduction by Ryan Grant (Evolution Publishing), pp. 93–108.
5 Cyprian, 2013 "On Jealousy and Envy," *The Complete Works of Saint Cyprian of Carthage* Translation Editing: Nicene Fathers [1885] Philip Campbell, Ed. Introduction by Ryan Grant (Evolution Publishing), pp. 161–170.
6 Cyprian, 2013 "On Mortality," *The Complete Works of Saint Cyprian of Carthage* Translation Editing: Nicene Fathers [1885] Philip Campbell, Ed. Introduction by Ryan Grant (Evolution Publishing,) pp. 117–130.
7 Cyprian, 2013 "On the Lapsed," *The Complete Works of Saint Cyprian of Carthage* Translation Editing: Nicene Fathers [1885] Philip Campbell, Ed. Introduction by Ryan Grant (Evolution Publishing), pp. 73–92.
8 *De Laude Martyrii* in Harper, Kyle, 2016 "Another Eyewitness to the Plague Described by Cyprian, with Notes on the "Persecution of Decius"," *Journal of Roman Archaeology* 29: pp. 473–476. https://doi.org/10.1017/S1047759400072263.
9 Eusebius Pamphilus, 1890 *Ecclesiastical History* Trans. A.C. McGiffert (Pantianos Classics).
10 Feldman, H. and Klenk, H.D., 1996 *Filoviruses Medical Biology- NCBI Bookshelf; Chapter 72—Filoviruses*. https://www.ncbi.nlm.nih.gov/books/NBK8129/. Accessed on 11/29/2020.
11 "The Flu—Votive Offering to Apollo, Salus and Aesculapius," 2018 *CoinTalk*. https://www.cointalk.com. Accessed on 2/26/2022.
12 Haldon, J., Elton, H., Huebner, S., et al., 2018 "Plagues, Climate Change, and the End of an Empire: A Response to Kyle Harper's *The Fate of Rome*," *History Compass/Wiley* Parts 1–3 [inclusive]. https://doi.org/10.1111/hic3.12508. https://doi.org/10.1111/hic3.12506. https://doi.org/10.1111/hic3.12507.
13 Hanson, J.W. and Ortman, S.G., 2017 "A Systematic Method for Estimating the Populations of Greek and Roman Settlements," *Journal of Roman Archaeology* 30: pp. 301–324.
14 Harper, Kyle, 2016 "Another Eyewitness to the Plague Described by Cyprian, with Notes on the "Persecution of Decius"," *Journal of Roman Archaeology* 29: pp. 473–476. https://doi.org/10.1017/S1047759400072263.
15 Harper, Kyle, 2015 "A Time to Die: Preliminary Notes on Seasonal Mortality in Late Antique Rome," in Laes, C. et al., eds., *Children and Family in Late Antiquity: Life, Death, and Interaction* (Leuven), pp. 15–34.
16 Harper, Kyle, 2017 *The Fate of Rome: Climate, Disease & the End of an Empire* (Princeton University Press).

17 Harper, Kyle, 2015 "Pandemics and Passages to Late antiquity: Rethinking the Plague of Cyprian c. 249–270," *Journal of Roman Archeology* 28: pp. 223–260.

18 Harper, Kyle. 2017 "Solving the Mystery of an Ancient Roman Plague," *The Atlantic*, 11/1/2017. https://www.theatlantic.com. Accessed on 4/13/2020.

19 *Historia Augusta, Lacus Curtius* Trans. David Magie, 1932. https://www.penelope.uchicago.edu. First Accessed on 4/20/2020.

20 Horgan, John, 2016 "Plague of Cyprian, 250–270 CE," *Ancient History Encyclopedia*. www.ancient.eu/article/992/plague-of-cyprian-250-270-ce/. Accessed on 04/16/2020.

21 Huebner, Sabine, 2021 "The 'Plague of Cyprian': A Revised View of the Origin and Spread of a 3rd-c. CE Pandemic." Published online by Cambridge University *Journal of Roman Archaeology* 34(1) (June): pp. 151–174.

22 Kelly, John Daniel, Wannier, Sarah Rae, et al., 2020 "The Impact of Different Types of Violence on Ebola Virus Transmission During the 2018–2020 Outbreak in the Democratic Republic of Congo," *Journal of Infectious Diseases* 222. https://doi.org/10.1093/infdis/jiaa163. Accessed on 3/18/2021.

23 Martinez, Micaela, 2018 "The Calendar of Epidemics: Seasonal Cycles of Infectious Diseases," *PLoS Pathogens* 14(11): e1007327. https://doi.org/10.1371/journal.ppat.1007327.

24 Martins-Filho, 2021 "Relationship Between Population Density and COVID-19 Incidence and Mortality Estimates: A County-Level Analysis," *Journal of Infection and Public Health*. https://doi.org/10.1016/j.jiph.2021.06.018. Accessed on 7/19 2022.

25 Mayo Clinic, 2019 "Ebola Virus and Marburg Virus." https://www.mayoclinic.org. Accessed on 04/15/2020.

26 McCormick, Michael, et al., 2012 "Climate Change During and After the Roman Empire: Reconstructing the Past from Scientific and Historical Evidence," *Journal of Interdisciplinary History* 43: pp. 169–220.

27 McNeill, William H., 1976 *Plagues and Peoples* (Anchor Books), pp. 104–108.

28 Miller, S.S., 1992 "R. Hanina bar Hama at Sepphoris," in Levine, L.I., ed., *The Galilee in Late Antiquity* (New York), pp. 175–200.

29 Paulus Orosius, 1964 *The Seven Books of History Against the Pagans* Trans. Roy Deferrari (Catholic University of America Press, Washington).

30 Pontius the Deacon, 2013 "The Life and Passion of Cyprian" The *Complete Works of Saint Cyprian of Carthage* Translation Editing: Nicene Fathers [1885] Philip Campbell, Ed. Introduction by Ryan Grant (Evolution Publishing), pp. 3–18.

31 Rathbone, D. and von Reden, S., 2015 "Mediterranean Grain Prices in Classical Antiquity," in Van der Spek, R.J. et al., eds., *A History of Market Performance: From Ancient Babylon to the Modern World* (London).

32 Rives, J.B., 1999 "The Decree of Decius and the Religion of Empire," Journal *of Roman Studies* 89: pp. 139–141.

33 Rohrbacher, David, 2013 "The Sources of the *Historia Augusta* Re-examined," *Histos* 7: pp. 146–180.

34 Thompson, Corinne, Baumgartner, Jennifer, et al., 2020 "Covid-19 Outbreak—New York City, February 29–June 1, 2020. Centers for Disease Control." https://doi.org/10.15585/mmwr.mm6946a2. Accessed on 7/19/22.

35 *Tractate Taanit*, "Ch.3: Regulations Concerning Occurrences on Account of Which Fast-Days Are Stopped, or Alarms Are Sounded. When Fasting on Account of Rain Is Stopped," pp. 48–52 and CH 3 (inclusive). *Jewish Virtual Library*. https://www.jewishvirtuallibrary.org. Accessed on 4/14/2020.

36 *Yerushalmi Horayot* 2 48b and *Yerushalmi Shab* 14 14c *Jerusalem Talmud*, 2010.

37 Zosimus, 2004 *The New History*, Taylor Anderson, Ed. (Odin's Library Classics).

CHAPTER 3

A Most Difficult Source and the Relevance of Climatic Circumstances

Our Initial Analysis of the Dionysius/Eusebius Account

Dionysius first appears in Eusebius' account when the former assumed the office of Bishop of Alexandria upon the death of the preceding office holder Heracles toward the end of 248 CE. From here on, Eusebius utilizes now lost letters of Dionysius. Dionysius' sole focus is, at first, on the subsequent religious persecution and martyrdom of Christians in the Roman Empire's leading African metropolis. By late 248 (with Philip still in power), anti-Christian rioting began in Egypt; the situation in Alexandria, in particular, began to deteriorate. As Dionysius related in Book VI Chapter 41, titled "The Martyrs in Alexandria…." "The persecution among us did not begin with the royal decree (fall 249 CE), but preceded it an entire year." The reference here is clearly to the single "royal decree" of Decius and *not* the Two Edicts of Emperor Valerian, whose later persecution of Christians is not yet a focus in Book VI of Eusebius' *Ecclesiastical History*. Book VI describes the period of late 248 to 250 CE as two years of religious and political violence and persecution in Alexandria with Christians often being the main targets. It all began in late 248/early 249 CE. Mobs attacked Christians in the street as well as in their homes—robbing, looting, torturing, and killing. Unlike the ensuing Decian Persecution, this was no official, organized, often effective, bureaucratic pressure campaign designed to coerce/reinforce

M. Orsag et al., *Interdisciplinary Insights from the Plague of Cyprian*, https://doi.org/10.1007/978-3-031-26094-0_3

tradition, cultural conformity, and political loyalty within the Roman Empire. Christians were instead set upon by mobs who tortured their victims if they didn't surrender property and join with their tormentors in "impious cries." Dionysius notes, however, "I know of no one unless someone possibly fell into their hands, who, up to this time, denied the Lord." In fact, seizure of property rather than enforcement of religious tradition seems to have been the major motive for the pagan attackers—"the more valuable property they took for themselves." Many Christians, including Dionysius apparently fled, "withdrew and went away" in response to the violent mob attacks, accepting "the spoiling of their goods" that they had left behind [23, p. 124 (6.41.1-8)].

Next, "... a sedition and civil war came upon the wretched people and turned their cruelty toward one another." This, in our view, likely refers to the attempted usurpation of Marcus Jotapianus against Philip and his brother and *Rector Orientis* Priscus in the Roman Empire's eastern provinces, which were, in the words of Zosimus, beset by "many disturbances." There were indeed a number of regional attempted usurpations against Emperor Philip the Arab, so many that, at one point, he considered resignation. None of the other usurpers, however, match up regionally or in terms of the scale of violence described by Dionysius except for the Antioch-based, and thus fairly nearby, Jotapianus. His rebellion spread in Syria, Cappadocia, and other eastern provinces and continued locally and sporadically there even after this death. Jotapianus got as far as minting coins before eventually being assassinated by his own troops. This unrest in Alexandria, we will argue presently with supporting evidence to come, apparently *concluded* in March (before Easter) at the latest, which appears too early in 249 CE to be linked to Decius' ultimately successful (by September 249 CE) spring rebellion against Philip. This timing later helps potentially situate, in our view, Dionysius' pair of relevant letters (to Hierax and to the brethren in Egypt) in late 248–249 CE—the period of and following the height of Jotapianus' rebellion. Angry Egyptian and Middle Eastern supporters of Jotapianus (incorrectly referred to by Zosimus as "Papianus") were probably motivated by resentment of heavy taxation by Philip and Priscus (the latter described by Zosimus as a man of "an intolerably evil disposition"). The increased tax revenue was designed to refill the imperial coffers after a series of wars, tribute payments, and the lavish Millennium Games (staged by Philip to celebrate the founding of Rome) had effectively emptied them. Before working up the courage to take on the riskier task of trying to get rid of their despised emperor and his family, the perhaps taxation-impoverished Alexandrian mobs may have first vented their anger and desire for loot on an unpopular and easier target at least loosely associated with Philip's unpopular clan—Christians. This, we believe, followed by expressing their desire for political "innovation" by supporting Jotapianus' regional sedition [23, pp. 124–126 (6.41–42); 40, pp. 6–7 (1.13–15)].

By contrast, the fall 249 CE accession of Decius, after his victory over and assassination of Philip at faraway Verona, seems to have initially produced little actual violence in Alexandria. "But presently, the change from that milder reign [of Philip] was announced to us...." This was the official start of the so-called Decian Persecution and its demands for public and legally recorded religious and political conformity through sacrifice. "For the decree arrived... All were truly affrighted. And many of the more eminent in their fear came forward immediately; others who were in the public service were drawn on by their official duties; others were urged on by their acquaintances... as their names were called, they approached the impure and impious sacrifices." This was an event during which Dionysius in Alexandria, like Cyprian in Carthage, also fled (yet again). Many other Christians, according to Dionysius, did the same, "Why need I speak of the multitude that wandered in the deserts and mountains and perished by hunger and thirst and cold and sickness and robbers and wild beasts?" The fact that this formal campaign occurred at this time is possibly an additional clue that pestilence was already impacting Roman Egypt and the Middle East. As Early Church scholar Bernard Green noted, "If the gods ignored the due worship of the city and still inflicted famine or disease or conflagration upon it, then it seemed more than likely that some people had alienated them." Under the circumstances, Christians had plausibly become an obvious and tempting [now official] target [7, p. 133; 23, pp. 124–126 (6.41–42)].

During the rest of Book VI, the Dionysius/Eusebius account retains its focus on martyrs and persecution; though there is also extensive discussion of a religious controversy within the church that arose directly out of the Decian Persecution. This concerned the so-called "lapsed Christians" who had sacrificed to the pagan gods during the campaign but then later repented. "What counsel then, brethren, do you give us concerning such persons?" The text at the beginning of Eusebius' Book VII continues in only a slightly broader vein and mentions that Decius' successor Gallus "stumbled on the same stone" [of expressing hostility to and persecuting Christians]; though, apparently, not to the same degree as Decius. As noted earlier, Decius' decree expired completely with his death. Fairly brief discussions of far less severe and systematic persecutions, respites from them and several schismatic controversies (including continuing fierce debate within the Church about how to handle Christians who had "lapsed" during the Decian Persecution) dominate the first nine Chapters of Book VII. The listing of Bishops of Rome mentioned in order in these sections—Cornelius, Lucius, Stephen, and finally Xystus also help firmly date this period of partially "restored peace," starting in approximately late 251 CE, and culminating fully and finally in the early, initially favorable, stages of Valerian's reign. It ended with the beginnings of Valerian's Persecution—the issuance of the first of the Two Edicts in 257 CE [23, pp. 126–132 (6.42–7.9)].

Dionysius continues his account with the reign of Emperor Valerian (r. 253–260 CE), in Chapters 10–11 and focuses intensely on the "persecution which prevailed so fiercely in his reign." Dionysius highlights the instigating role of the advisor to Valerian (and future, with his two sons, failed usurper) Macrianus. In Dionysius' view, it was Macrianus who turned the originally "mild and friendly" Valerian against Christians. The virulently anti-Christian Macrianus was strongly castigated by Dionysius.... "Wherefore, he became an enemy of his Catholic Church ..." Dionysius was exiled to Libya by Macrianus' ally, the Alexandrian prefect Mussius Aemilianus. Religious themes of martyrdom, miracles, holy relics and a relative respite from religious persecution during the reign of Emperor Gallienus (r. 260–268 CE), after Valerian's capture by the Persians, dominate Chapters 12–19. Gallienus became Emperor and was immediately more sympathetic toward the Christians, telling them at one point, "...no one may molest you...." This era is labeled the "Peace under Gallienus" by Dionysius/Eusebius. This section closes (Chapter 20) with a reference by Eusebius to an additional (many types of letters written by Dionysius—private, official to the Bishops of Rome, diaconal, etc., had already been discussed by Eusebius) collection of Dionysian letters. These were annual festal letters written by Dionysius during (and apparently prior to) Valerian's reign, "in which he sets forth a canon of eight years...." Dionysius argued therein that "...it is not proper to observe the Paschal feast until after the vernal equinox" [23, pp. 133–138 (7.10–20)]. As we shall see, these festal epistles, issued to the suffragan sees just after Epiphany and thus near the beginning of each year, can serve, in context, as a quite effective analytical dating mechanism within this relevant section of the *Ecclesiastical History*. This is particularly true if there is a strongly confirmed date to work backward or forward from, mathematically, apparent within the text.

Though, there are still, predictably with this crucial but difficult source, certain frustrations and less than perfect clarity overall. Are the "eight years" referring somehow to the reigns of Valerian and his son and co-Emperor, Gallienus? The fit isn't perfect. Does this description cover the entire span of Dionysius' festal letters or some fraction of it? When, within the temporal span of Dionysius' tenure as Bishop of Alexandria, did the practice of issuing Festal Letters originate/evolve? There is mention of a letter apparently dating to the early 250s CE that sounds very much like an early festal letter of Dionysius, but maddeningly is not clearly identified as such by Eusebius in Book VI, Chapter 46. Much of this is, to us, still rather unclear [23, p. 138 (7.20) and p. 128 (6.46.5)].

What is far more discernible is that, by declaring these letters as establishing holy writ, Eusebius is effectively inserting his knowledge of the evolution of Church doctrine in the fourth century. The First Council of Nicaea in 325 CE later universally completed the process seemingly started locally by Dionysius. The Council assigned to the Bishops of Alexandria a Christianity-wide role in denoting the date for Easter in an annual festal letter. This fully canonized this practice of the Bishops of Alexandria that had originated during Dionysius' tenure. Such festal letters of his announced the beginning of Lent and also contained "in addition a discussion of current ecclesiastical affairs or problems of Christian life and exhortations to observe the fast, to almsgiving and the reception of the sacraments." At this earlier stage, as noted above, Dionysius merely used these annual letters to locally push back the Paschal (Easter) Feast until after the vernal equinox. This would effectively, if somewhat roughly, center the seasonality of the peak of the Plague of Cyprian outbreak in Alexandria to the last 10–12 days of March or the month of April for any year following the years of the Valerianic Persecution. However, there is something of a complication. If our dating of 249 CE is correct, that year (likely if not certainly) preceded any full implementation of Dionysius' reform. Here, we must, for the sake of brevity and clarity, somewhat simplify, in multiple senses, a very complex matter (hopefully without distortion). The celebration of Easter was not yet consistent regionally, but largely had, albeit varying, linkage to the doubly astronomically determined and more complicated annual dating of the Jewish eight days of Passover. Ultimately though, the range of potential dates, in our context is not greatly altered.

The span being approximately from around March 21st to April 27th [23, pp. 130–136 (7.1, 7.10, 7.20, 7.11.1, 7.13), 30].

In Chapter 21 of Book VII, the testimony of Dionysius, as related by Eusebius, temporarily begins to shift away from more exclusively religiously oriented matters (persecutions, flight, martyrdom, heresy, etc.) to other less-specifically religious events ("Occurrences," "Pestilence," and a political "Reign") during the time period in which Dionysius was Bishop in Alexandria. Chapters 21," The Occurrences at Alexandria" and 22, "The Pestilence that Came Upon Them" and Chapter 23 "The Reign of Gallienus" present events related to the great city of Alexandria or the Roman Empire more broadly and in greater detail through Eusebius' recapitulation of Dionysius' eyewitness testimony or commentary. While a strict theological focus is no longer at the fore, the effects of these events on the Christian community is ever present. At first, this is done through a letter written by Dionysius to an Egyptian Bishop named Hierax. This letter again refers to, in our view, the period in late 248/early 249 CE after mob violence against Christians had abated. It clearly references a considerable span of time. We want to be very precise and clear here. In our view, the letter to Hierax very plausibly dates to late 248 to early 249 CE, but due to a possibly inserted (noted earlier) last section that references two different time periods and may (or may not) represent two different authorial "voices" and Huebner's analytically valid cautions (see all below) regarding Harper's earlier conclusions about the letter, we cannot (independently of our analysis of what follows in Chapters 21 and 22 of the *Ecclesiastical History*) conclusively date the letter to Hierax [23, pp. 107–149 (7.21.1–6,7.22.4,7.23)].

Dionysius seems to have returned to the city, but "sedition and war" and "insurrection" were about to break out. There is no mention yet of famine as there had been none earlier in Book VI Chapter 41. The language ("sedition and war"), the underlying chronology, order of events, and explanatory details here replicate that of Book VI Chapter 41, which definitely covers the year prior to the beginning of the Decian Persecutions. Plausibly, these ground the letter of Dionysius to Hierax in 248–249 CE chronologically, textually, and through our (admittedly potentially circular) analytical linkage to Marcus Jotapianus' attempted usurpation. As he demonstrated during later persecutions, Dionysius had a habit of fleeing Alexandria during violence directed at Christians. He is thus described as having "returned to Alexandria." After, he reveals in the slightly later second letter to the brethren in Egypt, "…they drove us out." This is, we believe, a reference to the anti-Christian mob violence described in Chapter 41 of Book VI, dating to late 248/early 249 CE. The brief respite in the disorders in Alexandria had allowed Dionysius to return, but the renewed outbreak of "sedition and war" in the city (associated, in our view, with the aforementioned rebellion of Marcus Jotapianus...) seems to have confined Dionysius in his house as going anywhere else in the disordered city—going "from Alexandria to Alexandria itself," as Dionysius somewhat poetically explained it, became nearly impossible. Again, the sequence seems to be the very same one that we have clearly dated to the end of Philip's and the beginning of Decius' reign [23, pp. 124–139 (6.41, 7.21.1–3, 7.22.4)].

In the letter to Hierax, Dionysius also gives an overview of prevailing recent climate conditions in a way that is remarkably in sync with what was evidenced by the *Baraita* appearing in the *Tractate Taanit* for the late 240s CE. Dionysius, a lifetime resident of Alexandria, seems to express astonishment while describing recent episodes of a dry waterless Nile followed by unpredictable sudden violent "overflow[s]." In ways similar to the problems of consilience regarding contemporaneous regional drought issues revealed in the *Tractate Taanit*, these descriptions also seem to, somewhat, bear out Harper's and Rathbone's characterization of a "more erratic regime of Nile flooding," at least for the 240s CE. Though this point also doesn't necessarily contradict the contentions of Haldon, Elton, Huebner et al., that there is no real proof of *overall decline* in Nile flood frequency or quality in the broader third century. Nor does it indicate that all the years of the 240s CE in Roman Egypt and the Middle East consisted of unbroken drought and failed floods. In this regard, the *History Compass* authors argue for six "normal or even good years," between 239 and 250 CE, by citing the work of the respected French papyrologist Danielle Bonneau, whose work Harper also cites. Though, in fairness to Harper, it must be pointed out that he actually used not Bonneau, but rather an evidentiary combination of Dionysius/Eusebius, the 13th Sibylline Oracle, the 1980s work of the Italian Egyptologist Casanova, several papyri (from different locations) and the research of Rathbone and von Reden for this narrower purpose related to the drought of the 240s CE in *The Fate of Rome*. Seemingly variable precipitation and flooding regimes—such periodic sudden "dry to wet" events and related Nile overflows would have been fueled, in large part, by increased volume from renewed rainfall in the Ethiopian highlands, the wellspring of the Blue Nile. The Ethiopian rainy season runs generally from June to September but can begin as early as March. All of these puzzle pieces of data and analytical conclusions collectively do seem to indicate a probably intermittent but quite damaging drought in the region in the 240s CE. It was, however, likely repeatedly broken, characteristically suddenly and violently, in ways that weren't always ideally supportive of a rebound in agricultural production. This broader climate pattern is clearly evidenced in both of the ancient literary sources that we have analyzed in detail—Dionysius/Eusebius and the *Tractate Taanit* [8, 11, pp. 132–133 and 332 (note 33); 23, pp. 138–139 (7.21.1–6, 7.22.4)].

Such sudden "dry to wet" events indeed seem to have been a repeated regional pattern during the 240s CE—annually most probably associated with the February–April and August–October windows as inferred from the mutually reinforcing combination of both Dionysius' account and the *Tractate Taanit*. But Dionysius' additional linkage, through biblical references concerning Moses changing water into blood, of the newly overflowing Nile to the violence potentially associated with Jotapianus' sedition is a further marker through which we can also attempt to link at least one of these "sudden dry to wet" episodes to the outbreak of the plague. Additionally, the evidence from the *Tractate Taanit* reinforces the notion, derived from Dionysius' description of events, that the dramatic return of wetter conditions and Nile flooding near Alexandria would not necessarily have alleviated food shortages during in the region. A famine is referenced in the *Ecclesiastical History* in the following letter to the brethren in Egypt that is the focus of Chapter 22 [23, pp. 107–149 (7.21–22)]. Dionysius' lengthy description in the letter to Hierax was clearly influenced by both Christian religious allegory and references to prevailing Greco-Roman medical concepts linking disease to polluted water and miasmas. He thus associates this suddenly renewed flow of water polluted with blood directly with the outbreak of the Plague of Cyprian, which followed the violence and the return of these suddenly wetter conditions; though non-believers and/or the less well-educated, in Dionysius' view, apparently failed to make such an association:

> And the river which flows by the city has sometimes seemed drier than the waterless desert, and more parched than that in which Israel, as they passed through it, so suffered for thirst that they cried out against Moses…
>
> Again, it has overflowed so greatly as to flood all the surrounding country, and the roads and the fields; threatening to bring back the deluge

> of water that occurred in the days of Noah. And it flows along, polluted always with blood and slaughter and drowning, as it became for Pharaoh through the agency of Moses, when he changed it into blood, and it stank.
>
> ... Or when can the air poisoned by these noxious exhalations become pure? For such vapors arise from the earth, and winds from the sea, and breezes from the river, and mists for the harbors, that the dews are, as it were, discharges from dead bodies putrefying in the elements around us.
>
> Yet, men wonder and cannot understand whence these continuous pestilences; whence these severe sicknesses [sic]; whence these deadly diseases of all kinds; whence this various and vast human destruction.... [23, p. 138 (7.21.5–9].

Huebner raises some, we believe, legitimate cautions about both Harper's aforementioned specific interpretations concerning the death toll from the plague in the city and his temporal placement and analysis of both this passage and what comes next—Dionysius/Eusebius' account regarding the depopulation of Alexandria: "...But the men from forty to seventy years of age were then so much more numerous that their number cannot now be filled out, even when those from fourteen to eighty years are enrolled and registered for the public allowance of food." It is indeed unclear "when" either "then" or "now" in the text really represent? Nor is it clear whether this is still Dionysius' "voice" or a later observation inserted by Eusebius? Do these passages conclusively date the events described or not? If so, when? To us, this all seems relatively inconclusive. Do the passages reflect multiple outbreaks in Alexandria during the plague period or merely indicate Dionysius' use of Christian religious allegory as a means to frame and interpret third-century events? We have come to no clear conclusions regarding these vexing questions. We have thus only firmly analytically interpreted the letter to Hierax in four more shallow/general senses: (1) The Plague of Cyprian was a contributing factor to multi-causal severe depopulation, over an unknown temporal span in the mid-third century, in the city of Alexandria. (2) The first passage is also entirely consistent with the regional "violent" drought-breaking described in the *Tractate Taanit.* (3) It is also consistent with our interpretation that such sudden dry to wet events were a repeated regional phenomenon during the relevant time period. 4). Finally, the relevant passage clearly indicates that the disease outbreak in Alexandria followed the breaking of the drought and the sudden, dramatic return of wetter conditions [15, p. 152; 23, pp. 138–139 (7.21.9–10)].

Next comes the letter to the brethren in Egypt, where we ultimately find ourselves on far firmer analytical ground in terms of conclusively dating the letter. Book VII, Chapter 22, "The Pestilence which came upon them" constitutes the key section dealing with the outbreak of the plague in the *Ecclesiastical History*. Again, Dionysius apparently recapitulates the events that he had previously described in Book VI, Chapter 41—"For many terrible things had happened already..." the beginning of the sequence is again the same as what was referenced then regarding late 248–249 CE. This replicated sequence begins with the driving out of Christians ("First, they drove us out...") with some Alexandrian Christians wandering in inhospitable environments and others being martyred. Civil war follows. The vague language being used in Book VII Chapter 22 again parallels that describing the anonymous mobs of "heathen" persecutors and thus harkens back to 248–249 CE. This, as opposed to the rather more specific persecutors such as Decius, Gallus, Valerian, Macrianus or members of Roman officialdom such as judges, centurions, governors, soldiers, tribunals, prefects, *frumentarii*, etc., that populate the Dionysian accounts of the more organized later Decian and Valerianic persecutions. A famine, contemporaneous and apparently interconnected with the sedition and civil war, is additionally referenced in Book VII, Chapter 22. Then, after a "very brief" interval (again directly parallel to the "we breathed for a little while" description appearing in Book VI, Chapter 41), the pestilence enveloped Alexandria. Though the violence in the city receded, the disease seemed to have built to a terribly destructive crescendo by late March/early April of 249 CE. Dionysius noted this in his letter (definitely *not* a festal letter as that would have been issued some months earlier) regarding the pestilential event. He indeed seems to have been in Alexandria for some of the worst of the plague. This prompted him to take the clearly rather unusual step of writing the second letter to the brethren; apparently in response to Alexandrian Christians who were suggesting the cancelation of the Easter celebrations [23, pp. 107–149 (6–7)].

The Paschal (Easter) festival was, after all, taking place at a time "that might not seem suitable...because of the multitude of the dead and dying." Eusebius' concluding commentary (passages 11–12 in Chapter 22) reference later (seven to nine) festal letters written by Dionysius. Both passages, however, are at least initially slightly ambiguous in terms of temporal placement. The latter (12), however, does reference an even *later* ("... he wrote again") Festal epistle of Dionysius to

Hermammon "describing at length the wickedness of Decius and his successors and mentioning the peace under Gallienus." This is probably the same Hermammon mentioned in the Dionysian letter referenced by Eusebius in Book VII, Chapter 10, which describes "Valerian and the Persecution under him." This broadly appears to order the outbreak of plague described in Chapter 22 in Alexandria *before* both the Decian and Valerianic persecutions in 249 CE and "the Peace under Gallienus." Nevertheless, the overall, recurringly referenced sequential pattern of events in that source, as well as the broader supporting evidence we have offered, indicates to us that the Plague of Cyprian clearly peaked in Alexandria by the spring of 249 CE. The POC's epidemiological beginnings in the city, unlike those for Carthage, aren't initially clear from the available ancient literary evidence, however, and they were probably unnoticed amid the Alexandrian welter of persecution, seditious violence, and famine. We will, however, attempt to deduce them later in Part III (by different means). As in Carthage, the pestilence seems to have displayed its peak during relatively cooler weather. As noted earlier, we will further address these and other interrelated issues (also disputed between the analyses of Harper and Huebner) of the origins and epidemiological directionality of the Plague of Cyprian in Part III. Our systems' synthesis conclusions here and there, are additionally, we believe, interconnected and mutually buttressing through the achievement of multidirectional and multidisciplinary consilience. Overall, our timeline is also a slight modification of Harper's original one in that we date the letter to the brethren Easter 249 as opposed to Easter 250 CE [12, p. 227; 23, pp. 133–139 (7.10,7.22, 7.21.6)].

Historiographical Background and Summary of Dr. Sabine Huebner's Analysis of Dionysius/Eusebius' Account

Dionysius' letter and Eusebius' later rendering of it have, however, been the subject of a continuing and intense decades-long multinational scholarly debate. One main point of dispute is the dating of Dionysius' pestilential letter to the brethren in Egypt. One alternative being 249–250 CE and other competing dates being 252 CE or, more often, over a decade later during the reign of Emperor Gallienus (the 260s CE). The first alternative of 249 CE effectively assumes (as we have in regard to the letter to Hierax) that Eusebius somehow edited, re-ordered, and/or even spliced together either single or multiple sources with his own later commentary. While inelegant, this solution to some of the confusing aspects of the account seems far more plausible to us. A number of modern specialists seem to acknowledge that such practices weren't uncommon in Eusebius' writing. There were a number of cases where he did these exact sorts of things (leading to confusion or even some direct contradictions). The later 252 CE or Gallienic alternatives of 262 or 263 CE seem to assume in our view, far less plausibly, a second unknown major persecution completely undocumented in any of the sources of the period and/or an even more inelegant and confusing reordering/splicing by Eusebius. Or, alternatively, a later pattern of events that, to a truly bizarre extent, exactly replicated those of 249–250 CE—something else that is totally unrecorded. A compelling example of some of the latest (February 2020) scholarship that touches upon the matter, from Dr. David DeVore of California State Polytechnic University—Pomona, aligns strongly with the overall dating conclusions (concerning 249–250 CE) advanced here by our team, by Kyle Harper, and earlier (in 1978) by the German historian Dr. Wolfgang Beinart. DeVore concludes that there are "serious problems with the Gallienic date." While DeVore's work is centered on the use of traditional formulas (*topoi*) of classical historiography in the work of "Dionysius/Eusebius" to promote Christianity and discredit Pagans, DeVore also implicitly or expressly accepts many of the key assumptions regarding the Plague of Cyprian contained in what he labels Harper's "pioneering" work. This includes the earlier dating of the Plague of Cyprian in Alexandria and the fact that the pestilence clearly "shocked" the Roman Empire. The Plague of Cyprian's impact was great enough, in DeVore's estimation, that it was "unlikely to have been forgotten by Eusebius' time in the early fourth century." In fairness and transparency to both Dr. Huebner and Dr. DeVore, however, it is notable that Dr. DeVore is listed in Dr. Sabine Huebner's 2021 "Acknowledgments" at the end of her *Journal of Roman Archaeology* article. It is possible that his views on the subject have evolved since he published what we noted here [3, 4, 6, 15, p. 171; 31, 33–35].

Huebner begins her analysis by acknowledging the complexity of the long-running debate over the dating of the two letters of Dionysius relating to the outbreak of the Plague of Cyprian in Alexandria. A central pillar of her argument for a Gallienic dating of the two key letters is.... "The sequence of the letters defended by most scholars also respects the order in which they are presented by Eusebius..." [15, p. 154]. Huebner, however, views Harper's narrowly based evidence (the two relevant letters of Dionysius as presented by Eusebius) and analysis as coming to an overly firm but highly dubious conclusion in dating the outbreak of the Plague of Cyprian in Alexandria to 249–250 CE. Harper solely follows, in Huebner's view, the analysis and interpretation of the historian Dr. Karl Strobel (without following his rather unorthodox interpretation that in fact there was no actual plague) and ignores a broader and more balanced approach to both the available source material and the relevant historiography. She later concludes that there is actually "no good evidence" for dating Dionysius' account of the pestilence in the letter to the brethren in Egypt "incontestably" to 249 CE as Harper does. Huebner, indeed, finds convincing evidence of an outbreak of the plague in Book VII Chapter 11 of the *Ecclesiastical History* that seems to date to the late 250s CE. She concludes that no evidence of the plague in Alexandria actually precede this much later date [15, pp. 152–156].

Huebner also strongly challenges Harper's interpretation of the passages regarding climate conditions, plague outbreaks, and the depopulation of Alexandria discussed in the Letter to Hierax. We freely acknowledge, as noted above, that Huebner raises multiple legitimate cautions about Harper's interpretation of those passages regarding the depopulation of Alexandria that also caused us to become far less ambitious about reaching our own analytical conclusions regarding these sections of Dionysius' letter to Hierax than we had previously been earlier in the evolution of our research. In fact, this letter to Hierax, rather than that to the brethren in Egypt, is central to her overall analysis. Through use of a table, she also demonstrates the preponderance of scholars who have dated both accounts to the Gallienic period. The references to sedition and civil war she explains as references to the attempted usurpation of the two sons of Macrianus (the Macrianni Brothers) against Gallienus in 260–261 CE (and its local Alexandrian/Egyptian coda involving the prefect Mussius Aemilianus that Huebner argues concluded only in the spring of 262 CE). Here, Huebner cites, while noting the inherent analytical difficulties involving the former, the *Historia Augusta* and papyrological evidence. Having thus collectively reviewed the evidence and interpreted it, Huebner asserts that Dionysius had "spent the years 249–251 CE in hiding and the span of time from 257 to 260 CE in exile." It was only after that period that Alexandria was afflicted with war and famine. Huebner deduces, through this timeline, an analytical linkage to the revolt of the Macrianni and her aforementioned analysis of the letter to Hierax that, in her view, clearly dates it to the Gallienic period. She accordingly dates the letter to the brethren, which she discusses in a far more limited manner, detailing the pestilence "most likely" to Easter 263 CE. This conclusion of Huebner's thus challenges Harper's overall plague timeline at its beginnings but also at its terminus. Harper, as noted, had dated the Plague of Cyprian, as a whole, to 249–262 CE. The cascading effects in terms of the analytical viability of Harper's origin evidence and conclusions would also be quite significant if Huebner's, erudite, apparently internally consistent and inherently plausible interpretation of Dionysius' relevant letters were to prove analytically unassailable [15, pp. 152–156].

Our Rebuttal to Dr. Sabine Huebner's Conclusions Concerning the Dating of "The Pestilence Which Came Upon Them"

Dionysius' letter to Hierax's aforementioned passages 5 and 6 clearly describe prevailing climate patterns (repeated sudden dry to wet episodes). Huebner's analysis of the concluding sections (7–10) raised some real doubts in our minds about Harper's interpretation of the letter and also caused us to change our original interpretation. We would also conclude, however, that she potentially over-interprets these passages in support of the 263 CE dating of the following account contained in Chapter 22. Passages 7–10 of the letter to Hierax contain some strong elements of metaphor and religious allegory (references to Eden and Geon for instance). Though this, in and of itself, is not terribly problematic. The interpretive veil of Greco-Roman medical humourism, very strongly evidenced in these sections, is an additional factor that makes parts of these passages difficult to interpret definitively. While ancient observers could and did make highly detailed observations of pathologies, how they conceptualized those observations, from a modern point of view, is not entirely clear. Huebner, for example, interprets the use of the plural in passage 9 to definitively signal multiple prior waves of the pestilence. To us, this appears far from certain. This passage could alternatively reference a single months-long outbreak. Ancient sources such as Orosius and *De Laude Martyrii* also, for example, use the plural "diseases," when discussing what were clearly singular outbreak events. In another context, later in the *Ecclesiastical History*, so does Eusebius. This ancient linguistic phenomenon may stem from ancient observers (whether trained Greco-Roman physicians or not) linking different observed pathologies conceptually to different humours/underlying causes. As noted earlier too, passages 9 and 10 are very confusing temporally and seem, to us at least, to have been potentially spliced in later by Eusebius (whether from later writings of Dionysius or his own observations). Since the severe pestilence described in Chapter 22 would certainly have been a major contributor to the Alexandrian depopulation described in Chapter 21's concluding section, it seems, in a certain sense, illogical to assume otherwise [11, p. 140; 22, p. 316 (7.22); 23, pp. 138–180 (7.21.1–10, 9.8.1)]. Again though, we freely admit the very tentative and uncertain nature of our conclusions regarding the (independent) dating of Dionysius' letter to Hierax.

Leaving the confusing temporal uncertainties of the letter to Hierax behind, we now move to the beginning sections of Chapter 22 and Dionysius' absolutely crucial, in context, letter to the brethren in Egypt, which describes the relevant outbreak of disease in Alexandria. Passages 1–5 link the outbreak of "pestilential disease" to persecution, flight, famine, and war. We have argued earlier that this sequence clearly harkens back to the events of 248–249 CE described by Eusebius (and Dionysius) in Book VI. Some of the phrasing used in this point in Book VII, "For many terrible things have happened already..." seems to indeed suggest such recapitulation. Huebner, by contrast, argues that the events referenced are from the Persecution of Valerian in the late 250s CE and the later attempted usurpation of the Macrianni (Macrianus and his two sons) against Gallienus in 260–261 CE [23, p. 139 (7.22.1–5)]. The key differentiating clue here, we believe, lies in Dionysius' description of the war in passage 5 (which immediately follows a description of prior persecution and anti-Christian violence in passage 4):

> After these things war and famine followed, which we endured in common with the heathen. But we bore alone those things which they afflicted us, and at the same time we experienced also the effects of what they inflicted and suffered from one another; and again, we rejoiced in the peace of Christ which he gave to us alone [23, p. 139 (7.22.5)].

This indicates that the war itself, unlike the persecutions that had preceded it, seemed to be disconnected from sectarian tensions in Alexandria. Christians were merely suffering bystanders. That Dionysius would have thus described a "sedition and war" [23, p. 138 (7.21.1)] inspired by the Macrianni or their adherents (like Mussius Aemilianus) seems exceptionally unlikely indeed given his earlier (and later) references, replete with condemnatory biblical quotations, to the persecutor Macrianus, his persecution-instigating influence on Valerian and this attempted usurpation in Book VII Chapter 10:

> Wherefore he became an enemy of his Catholic Church, and alienated and estranged himself from the compassion of God, and fled as far as possible from his salvation. In this, he showed the truth of his own name.
>
> ...For Valerian, being instigated to such acts by this man, was given over to insults and reproaches [a reference to the emperor's later Persian captivity] ...
>
> But this man [Macrianus] madly desired the kingdom though unworthy of it, and being unable to put the royal garment on his crippled body, set forward his two sons to bear their father's sins. For concerning them, the declaration which God spoke was plain, "Visiting the iniquities of the fathers upon the children unto the third and fourth generation of them that hate me" [Exodus 20:5].
>
> For heaping on the heads of his sons his own evil desires, in which he had met with success, he wiped off upon them his own wickedness and hatred toward God [23, p. 133 (7.10.5–9)].

...and from Book VII Chapter 23 describing the attempt of Macrianus and his sons to depose Gallienus:

> 1... Then he [Macrianus], having betrayed one of the emperors that preceded him and made war on the other, perished with his whole family speedily and utterly...
>
> 2. For according to the word spoken by the Prophet Isaiah, "Behold the things from the beginning have come to pass, and new things shall now arise." (Isaiah 42:9). For as a cloud passing over the sun's rays and obscuring them for a little time hides it and appears in its place; but when the cloud has passed by or is dissipated, the sun which had risen before appears again; so, Macrianus who put himself forward and approached the existing empire of Gallienus, is not, since he never was. But the other is just as he was [23, p. 140 (7.23.1–2)].

An identical analytical verdict would indeed also apply to the subsequent revolt of the similarly virulently anti-Christian Prefect of Egypt Lucius Mussius Aemilianus who had, after all, tried Dionysius in a quite hostile manner and exiled him to Libya during the Valerianic Persecution [23, pp. 133–134 (7.11.3–14)].

While surviving evidence regarding Marcus Jotapianus is somewhat limited, his aforementioned rebellion seems to fit the description of a largely non-sectarian sedition and civil war far better. It also seems to have been motivated, like others under Philip, not *primarily* by sectarian issues but rather by resentment over high taxes. Jotapianus claimed descent from Alexander the Great and Emperor Alexander Severus (the latter an Emperor generally tolerant of Christianity). Eusebius, in fact, notes in Book VI Chapter 28 that Alexander's "household... contained many believers." This revolt seems to be a much better match to the language describing a rebellion in which the pagans primarily "... suffered from one another" (Book VII) or (nearly identically described) "turned their cruelty toward one another" [the latter passage from Book VI and dating to late 248–249 CE] than the later revolts of the virulently anti-Christian Macrianni or Mussius Aemilianus [16, 20, 23, pp. 120–139 (7.28–7.22.5)].

Chapter 22's concluding section, passages 11–12 together, as well as the following Chapter 23 also, in our view, collectively create a series of insuperable problems for Huebner's 263 CE dating regarding Dionysius' pestilential letter to the brethren:

> 11. After this epistle, when peace had been restored to the city, he [Dionysius] wrote another festal letter to the brethren in Egypt, and again several others besides this. And there is a certain one extant On the Sabbath and another On Exercise.
>
> 12. Moreover, he wrote again an epistle to Hermammon and the brethren in Egypt, describing at length the wickedness of Decius and his successors and mentioning the peace under Gallienus [23, p. 139 (7.22.11–12)].

Chapter 23 of Book VII, not discussed in Huebner's 2021 article, in fact quotes directly from that very last (also clearly festal) letter of Dionysius that followed a number of years (denoted by the number of annual festal letters, 7–9) *after* the outbreak of pestilence in Alexandria. As Eusebius notes:

> 1. But there is nothing like hearing his [Dionysius'] own words, which are as follows:

> ... But Gallienus was proclaimed and universally acknowledged at once an old Emperor and a new, being before them and continuing after them.
> 2... [There follows the aforementioned section condemning Macrianus]
> 3. And his kingdom, as if it had cast aside old age and had been purified of the former wickedness [Valerian's and Macrianus' persecutions of Christians] now blossoms out more vigorously, and is seen and heard farther, and extends in all directions.
> 4. He then indicates the time he wrote this in the following words:
> It occurs to me again to review the days of the imperial years. For I perceive that those most impious men, though they have been famous, yet in a short time have become nameless. But the holier and more godly prince, having passed the seventh year, is now completing the ninth, in which we shall keep the feast [23, p. 140 (7.23.1–4)].

This seems to date this final later festal letter (indeed it is the last festal letter of Dionysius mentioned in the *Ecclesiastical History*) firmly to very early (probably January) 262 CE—problematically more than a year *before* Huebner's Easter 263 CE dating of the pestilence itself. This seems to run directly counter to the aforementioned core of Huebner's argument that Eusebius' sequence of presentation of the letters should be analytically respected in a chronological sense. Earlier Gallienic dating—260–262 CE, in light of the interconnectedness of the end of Book VII Chapter 22 and the beginning of Chapter 23, also seems to run into insuperable timeline issues regarding the revolts of the Macrianni and Mussius Aemilianus. *Additionally, we now have our firm temporal endpoint.* The next task is thus to determine, with this reference point, when, from a Christian perspective, "peace had been restored to the city." Independently, this could, of course, very plausibly reference this renewed "Peace under Gallienus" and this would be consistent with Huebner's dating of 263 CE. There is, however, a serious deeper analytical problem with such an inference. Huebner herself dates Dionysius' death to 264 or 265 CE. In addition to the, in context, seemingly firm and highly problematic dating revealed by the final festal letter mentioned in CH 22, it seems essentially impossible that Dionysius would have written seven to nine additional annual Easter Letters between 264 and (at the outside) 265 CE. Eusebius' use of the word "extant" in this specific instance strongly implies that one or more of Dionysius' annual Easter letters from the referenced period may have been lost by the time that Eusebius compiled the evidence for and wrote the *Ecclesiastical History*—in roughly the first third of the fourth century [23, pp. 139–140 (7.22–7.23); 15, p. 156].

The annual nature and religious importance (including retroactively) of the Easter letters would have made the absence of any, in the archive of Dionysius' letters, obvious to Eusebius—even over a half-century later. This inclines us toward a total of approximately eight–nine festal letters to be counted backward; that lands us, temporally, in 253–254 CE. Does this analytical fit make sense in a way that produces consilience?

Our analysis, starting from 249 CE, as noted, contrastingly references the period of progressively restored "peace following the [Decian] persecution" detailed in Book VII Chapters 1–5. That span broadly stretched from the middle part of 251 CE (following Decius' death at Abritus) to the beginning of 257 CE (prior to the issuance of the first of Valerian's Two Edicts). Does this fit analytically with what appears in Chapter 22 passages 11–12 and the additional information contained in Chapter 23? It does, we believe, with near perfect multidirectional temporal and analytical consilience. Dionysius' account of the pestilence, in our view, definitely describes events from the spring of 249 CE. Our reasoning from there is as follows; Dionysius would flee the Decian Persecution later that year. Peace would definitely not have been considered "restored" during January or even the spring of 250 CE (the Decian Persecution was at its height and Dionysius had fled Alexandria). The first half of 251 CE also seems a poor choice as Decius, despite the campaign's lost momentum, would still have been in power and the "royal decree" still in legal force. Given the general perception of anti-Christian hostility conveyed regarding Gallus, 252 CE also doesn't seem a terribly compelling choice (though not totally implausible). Emperor Aemilianus' (r.253 CE) brief and insecure reign isn't even mentioned in the *Ecclesiastical History*. The beginning stage of Valerian's reign, with its "manifest friendliness and hospitality" from the new emperor (253–254 CE) seems, however, the far most likely match. Contrastingly, the math, from that temporal range, works perfectly. Eight–nine annual festal letters following the restoration of peace and concluding with the aforementioned epistle to Hermammon, at the end of Chapter 22, ordering the events of the era and concluding with the "mentioning" of the "Peace under Gallienus." A description of that period appears immediately following in Chapter 23. There were, in fact, two different mathematical combinations of starting dates and totals of festal letters that landed us precisely in 262 CE. Chapter 20 of Book VII also provides confirmation that Dionysius wrote festal letters during this period including "while the persecution [under Valerian] was still prevailing." The reference to the "canon of eight years" in relation to the festal letters also contained in Chapter 20 is also very analytically tempting to more decisively interpret in this regard, but the phrasing still seems too ambiguous and confusing for us to actually do so [7, pp. 107–149 (6–7)].

For us, all this evidence and analysis collectively, perhaps once again, "incontestably" (to borrow Harper's earlier phrasing) dates Dionysius' letter to the brethren and the outbreak of the Plague of Cyprian in Alexandria to 249 CE. His original dating of the letter of Hierax to 249 CE and the letter to the brethren to 250 CE could also be generally accommodated by the overall textual analysis that we have produced here. We believe, however, that such a direct temporal overlap between the height of the pestilence in Alexandria and the height of the Decian Persecution would have been very likely more directly noted by a Christian source like Dionysius/Eusebius. Since it wasn't, that dating seems less likely. Whether later waves of the Plague of Cyprian impacted Alexandria (and we find Huebner's affirmative analysis in that regard quite compelling) is also, in this narrower context, irrelevant. This conclusion that there indeed was, almost certainly, a major plague outbreak in Alexandria in 249 CE has, in our determination, a very strong buttressing effect, on Harper's theory of origin, overall timeline and proffered epidemiological directionality. Our complex evidentiary reasoning here will, undoubtedly, face scholarly scrutiny. In the meantime, however, while some minor caveats, particularly in regard to the specifics of translation, may yet lurk, this seemingly solidly evidenced analytical conclusion now allows us, once again, to move forward with our broader modeling of the Plague of Cyprian as a whole.

So far, we haven't produced an estimate of how long the "burn and build" pattern of the Plague of Cyprian took in Alexandria's case, but, at least by implication, the latter possibly showed a similar or perhaps slightly more rapid epidemiological (outbreak to peak) timeline to the 6–7 months that we have estimated for Carthage (given the aforementioned evidence in Cyprian's account). If we, not illogically from the existing timeline evidence, assume a post-peak period of disease in both cities in late spring/early summer with a tailing off of the outbreak with abatement in the mid-to-late summer, the timeline in the Carthaginian and Alexandrian outbreaks very closely matches Symeon's described seasonality. Even if we attempt to avoid the potential circularity inherent in such an assumption through being slightly less specific for dating the end of the Carthaginian epidemic, for example, the timelines still match *as far as can be determined*. In Parts II and III, we will examine both timelines in much greater detail and also estimate the Alexandrian initial outbreak to peak timeline at somewhere around 5–7 months. If so, why weren't the epidemiological timelines more different? Alexandria was a much larger city than Cyprian's Carthage, with a population of around 411,000 people and an estimated population density of 422/ha (as opposed to 298/ha) for Carthage. Alexandria was also closer to the Kingdom of Aksum, and more easily accessible from, places like Berenice, Myos Hormos, and Luxor that were directly situated on the key Red Sea, Nile, and overland trade routes connecting the Roman Empire to Aksum and beyond. Alexandria was also more of a commercial, cultural, and administrative hub in general [10, p. 319]. It was, in many ways, the third-century Roman Empire's second city after Rome itself. The bloody violence and the famine plaguing Alexandria in 249 CE could have, as noted in an epidemiological sense, also quite possibly contributed to the spread of the disease, weakened the population, and accelerated the plague's epidemiological timeline.

Bringing Harper's implication of a filovirus as the pathogenic cause of the Plague of Cyprian into the equation, modern outbreaks (of our clearly included African filovirus suspects) in places such as the Democratic Republic of the Congo (Ebola) and Angola (Marburg) have demonstrated both generalized/perceived and scientifically documented associations between violence and filovirus infection dynamics, as measured by contagiousness (R0/Re/Rt, etc.). These increases have been meticulously documented in relation to a variety of specific statistical patterns/types/levels of violence by modern epidemiological researchers. Some of the modern suspected amplifying causes of this twentieth-century epidemiological phenomenon concerning the effects of violence (such as disruption of contact tracing, inability to carry out vaccinations, difficulties in medical response mobility due to landmines, or violence-forced evacuation of international medical teams) clearly wouldn't apply in an ancient Roman setting like that of third-century Alexandria. Others, specifically associated with such well-documented perils of filovirus transmission via infected blood, "unsafe" burials or body disposals and overtaxed and disrupted medical systems serving as epidemiological *foci* due to unsafe procedures, imperfect disinfection practices (including unsanitary utilization, improper use, or unauthorized reuse of medical equipment) clearly would. Similarly, "*ville morte* events" or, conversely, the potential trapping of an infected person in the community as opposed to allowing for their isolation (much as Dionysius was trapped in his house in Alexandria) would clearly also *have* direct parallels in a violence-disrupted ancient third-century Roman city. Yet despite such evidence of greater violence and disruption and Alexandria's larger size and likely higher population density, the outbreak timelines between the two seem to have remained relatively similar with Alexandria's being, at most, only slightly faster. Why? We will examine this issue in Part III as well [7, pp. 138–139 (7.21.9–10); 18, 21, 23, pp. 140–141].

As noted, Harper concludes that the combination of violence, famine, and the Plague of Cyprian (most of all) may have essentially depopulated the city of Alexandria by nearly two-thirds due to the devastating plague and the other earlier sources of mortality and the de-urbanizing flight from all of them [7, pp. 140–141]. Though a degree of hyperbole may certainly exist, multiple waves of pestilence between 249 and 259 CE may figure into the equation. Huebner's aforementioned critical commentary, however, provides an additional note of caution. Yet, Eusebius' aforementioned recapitulation of Dionysius apocalyptic account nevertheless states:

> ...why this great city no longer contains as many inhabitants, from tender infants to those most advanced in life, as it formerly contained of those whom it called hearty old men. But the men from forty to seventy years of age were then so much more numerous that their number cannot now be filled out, even when those from fourteen to eighty years are enrolled and registered for the public allowance of food.
>
> And the youngest in appearance have become, as it were, of equal age with those who formerly were the oldest. But though they see the race of men thus constantly diminishing and wasting away, and though their complete destruction is increasing and advancing, they do not tremble [23, pp. 138–139 (7.21.9–10)].

While not fully transformed into a *ville morte* itself, Alexandria reportedly suffered massive multi-causal depopulation, with the plague being a very likely significant contributor (whether in one or multiple waves between the late 240s and 259 CE). The disease would probably have also spread locally in the near term via *ville morte*-type events, with de-urbanizing flight, to areas surrounding the city. In the longer term though, through exoduses and the resulting decreases in population densities and other resulting changes in behaviors, these relocations would have eventually subsequently resulted in decreased Re/t contagiousness values. In addition to key matters of chronology regarding the Plague of Cyprian and Alexandria, however, the overall synergy between Dionysius/Eusebius and the *Tractate Taanit* in regard to the pattern of sudden "dry to wet" drought-breaking regional climate patterns in the 240s CE and subsequent outbreaks of infectious disease (weaker but still present in the case of the *Tractate Taanit*) is crucially relevant and highly suggestive as well. This is because drier climate patterns being broken by a sudden return of wetter conditions, historically and in modern times, are associated with destructive outbreaks of viral hemorrhagic disease. The work of Dr. Rodolfo Acuna-Soto of the Universidad Nacional Autonoma de Mexico has shown this connection, for instance, between this specific dry to suddenly wet climatic pattern and repeated outbreaks of "cocoliztli" (pestilence)—by undetermined species of viral hemorrhagic fevers (VHFs), the broader group of RNA viruses that includes the filoviruses—in the eighth, ninth, and sixteenth centuries in Mexico. These plagues collectively killed upwards of 15 million people. While Acuna-Soto's case is somewhat more inferential for the earlier Mexican outbreaks in the eighth and ninth centuries, his evidence for the repeated outbreaks in the 1500s (most notably in 1545 and 1576–1580) is absolutely conclusive. "Tree ring reconstruction of rainfall patterns over Central Mexico" shows the VHF outbreaks occurring consistently in "years of abundant rain embedded in the midst of a sixteenth-century megadrought." Kyle Harper, in his 2016 *Journal of Roman Archeology* article "Pandemics and Passages to Late antiquity", also referenced Dr. Acuna-Soto's work [1, 2, 12]. A great deal of additional more precisely applicable and easily accessible scientific scholarship than Acuna-Soto's 2008 research on the *cocoliztli*, much of it also more recent and therefore not used by Harper in 2016–2017, further improves the ancient–modern evidentiary overlay, and thus tightens the pathogenic-suspect ring even further.

In modern-day Sub-Saharan Africa, NASA (National Aeronautics and Space Administration) satellite imagery is being used to search for a strikingly similar sudden "dry to wet" climatically linked VHF outbreak pattern, through calculation of differences in the normalized difference vegetative index (NDVI) value over time. NDVI shows the relative "greenness density" of a particular geographic location and is a good indicator of drought. Analyses of changes in NDVI over time can be used to illustrate the rapid recovery of vegetative growth from drought, or vice versa. The tropical rainforest biome of central Africa, with high annual and nearly seasonally continuous rainfall (and consequently high NDVI [there is also a new improved measurement model known as the enhanced vegetative index or EVI]), is significantly tied to VHF outbreaks in humans, nonhuman primates, and potential bat hosts. However, in Sub-Saharan African regions outside of this central African niche, greater variation in climatic seasonality occurs and VHF outbreaks in these regions are tied to dramatic climatic shifts from dry to rainy seasons. Among the biogeographically indicated filoviruses, Ebola's climate-linked outbreak pattern is more recurring, regular, and cyclic. Marburg's dry to wet climate association is also quite strong but far more sporadic in terms of the frequency of outbreaks. The mechanism at work in these similarly climatically patterned "dry to wet" African filovirus cases are the precipitation dynamics toward the end of the Sub-Saharan dry season, especially when a sharply drier period is followed by a sudden return to rain. This dynamic for Marburg, in turn, is presumably tied to viral dynamics in the known reservoir host, the Egyptian rousette bat (ERB), whose diet, overall body condition, and ability to give birth twice a year are dependent upon the rain. That this process works for Ebola as well is clearly noted, but not as well understood—yet. These meteorological events are informally known among twenty-first-century epidemiologists specializing in these deadly African diseases as "Ebola-triggering weather patterns," (though, as noted, the Marburg virus has shown a similar pattern to that of the more regularly occurring Ebola outbreaks). A similar approach has also been used to analyze outbreaks of another "sudden dry to wet" potential VHF Plague of Cyprian suspect; the case of that suspect pathogen—Rift Valley fever (RVF) will be dealt with below [10, 14, 19, 26–28, 36, 37].

Given this powerful additional evidence that considerably bolsters Harper's already strong circumstantial case for a VHF, and leaving aside the clearly applicable members of the *filoviridae* (Ebola and Marburg) for the moment, we now extend the sort of differential analysis conducted by Harper in terms of the Plague of Cyprian's pathology and epidemiology to all the known (approximately sixteen) viral hemorrhagic fevers that are mostly spread across the taxonomic families of the *arenaviridae*, *bunyaviridae*, and *flaviviridae*. In fact, Dr. Harper had already addressed some of these cases in his original differential analysis that appeared in *The Fate of Rome*. We merely had to apply his methods and expand the analysis specifically to where the interdisciplinary consilience of ancient and modern evidence clearly and compellingly pointed—the viral hemorrhagic fevers. On the basis of "not matching" geographical dissemination and/or known counter-indicated seasonality, (such as that of members of the mosquito-borne *flaviviridae)*, we were quickly able to eliminate all but a few of the non-filovirus VHFs as potential causes of the Plague of Cyprian. Most VHFs, though not all, are tropical (see Map 3.1). We used modern case fatality rates (CFRs) cautiously in our differential analysis here as corroborating evidence only. Modern medical care and potential variation in the basic pathogen biological components in relation to CFRs in the lineage ancestors of modern diseases are the dual roots of this caution. Mosquito-borne dengue with its characteristic rash and similarly vectored yellow fever with its characteristic jaundice-induced skin coloration have applicable biogeography but poorly matching pathologies and CFRs that are far too low. Tick-borne Kyasanur Forest disease is also eliminated by a combination of seasonality, displaying generally rural epidemiology and the disease's primarily East Indian biogeography. East Asian biogeography and recent molecular-phylogenetic emergence eliminate the SFTS virus along with, once again, a quite low CFR. South American biogeography decisively eliminates the Chapare, Guaranito, Junín, Machupo, and Sabia viruses. The Middle Eastern Alkhurma and the Russian Omsk hemorrhagic fevers are tick- or tick-and-rodent borne, primarily rural in epidemiology, lack human-to-human transmission, and have never caused large outbreaks. The very rare, highly lethal, and poorly understood Lujo virus was also eliminated (very strongly if not absolutely) on the basis of strong circumstantial evidence of very recent emergence and biogeography. Biogeographically included hantavirus HFRS was eliminated due to the lack of a human-to-human epidemiological dissemination pathway (relying instead on dissemination via rodent droppings). The disease is widely endemic but doesn't generally cause large outbreaks. Even the most virulent forms, such as the Dobrava-Belgrade virus, have CFRs of less than 15%.

Two other non-filovirus VHFs required a bit of a more detailed differential examination. The first of these initially promising suspects is the rodent-borne Lassa fever (of the arenavirus family). In his own differential analysis in *The Fate of Rome*, Harper dismissed Lassa (though given our enhanced case for a VHF, we decided to give it a more thorough second look). Harper's dismissal was on the solid grounds that the pathology and epidemiology of the Plague of Cyprian, through such features as the "speed of diffusion" and the fears of the sick and the corpses of the dead, overall, point to a "human-to-human" disease transmission that members of the *arenaviridae*, like Lassa, lack [11, p. 143]. Additionally, the geographical dissemination of this fever is fairly limited and centered

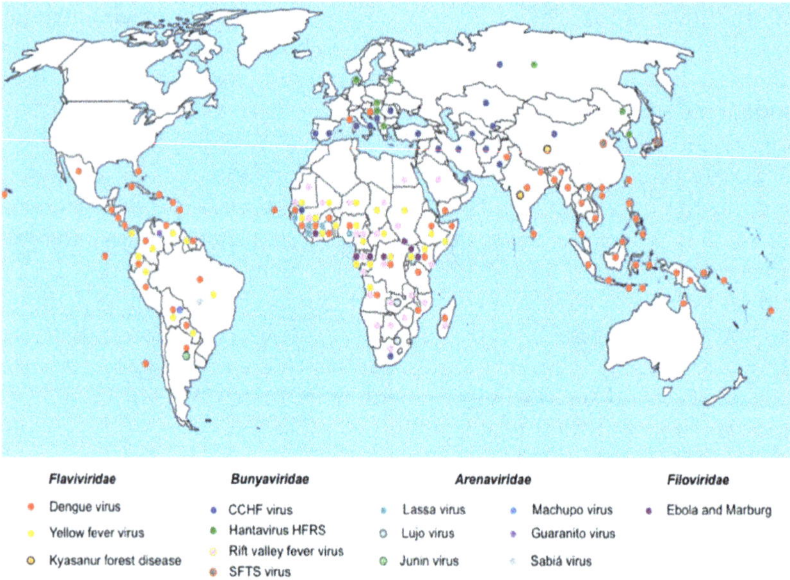

Map 3.1 Geographical Distribution of Viral Hemorrhagic Fevers. Map courtesy of Dr. Juan Carlos Zapata and PLoS.org 2014 [39]

entirely within modern West Africa (in an east to west line stretching from the modern nations of Nigeria through Guinea). While this level of biogeographical distance doesn't necessarily entirely rule Lassa out as a suspect, it would be far more logical that the Roman province of Mauritania would have been the heavily impacted starting point of the Plague of Cyprian for the empire, rather than Egypt—if Lassa fever had caused the plague. None of the ancient evidence even remotely implies that possibility. Lassa also has a mortality rate that is far too low, and its pathology, in a variety of ways, is also a very poor fit for the Plague of Cyprian.

One other more complicated case not addressed by Harper, but noted above, is Rift Valley fever (RVF). The geography and the strong association with sudden dry to wet patterns make it a potentially very strong pathogenic candidate. RVF is a rather unusual and highly epidemiologically opportunistic member of the phlebovirus family. It can be spread by zoonotic transmission or can be mosquito-borne. RVF is primarily a disease of cattle that can spread to humans. While its biogeographical dissemination in northeastern Africa is a perfect fit with Harper's and our Plague of Cyprian origin theories, and the ancient Aksumites raised cattle, Rift Valley fever (like Lassa) also lacks the highly pathogenic nature displayed by the Plague of Cyprian and does not display human-to-human transmission. The lack of multiple mentions in the ancient accounts of large numbers of dead cattle or other livestock such as sheep, pigs, or goats can also be counted against RVF in this context. One ancient Christian source, the fifth-century writer Paulus Orosius, indirectly mentions dead "beasts" and "infected grass" in an allegorical way that might imply dead livestock but only as part of an extended religious (again also clearly influenced by Greco-Roman medicine's conception of miasmas), and characteristically polemical, metaphor comparing the biblical plagues suffered by Old Testament Egypt and the pestilences suffered by pagan Rome. Orosius also claims that the Plague of Cyprian was an additional vengeance, visited on the city of Rome during the reign of Gallus and Volusianus, after the "early" death of their imperial predecessor—the persecutor of Christians—Emperor Trajanus Decius. This "seventh plague," as labeled by Orosius, also arose "from the poisoning of the air" and "also poisoned the lakes." Orosius thus is clearly engaging in anti-pagan religious rhetoric and (indirectly) again demonstrating the contemporary influence of Greco-Roman medical concepts linking disease to polluted water and miasmas as opposed to giving an account that can, analytically, convincingly be used to tie the Plague of Cyprian to RVF. Orosius may also have, given the specifics of the earlier similarly Christian Dionysius/Eusebius account, been utilizing and expounding upon (in a rather exaggerated manner), some version or portion of it. Like Lassa, RVF is also quite discriminating (pregnant women and their unborn children are particularly affected) in its pathology—unlike the indiscriminate Plague of Cyprian. No ancient account mentions anything along those lines [22, p. 327 (7.27)]. It is important to note that one potentially biogeographically included VHF pathogenic suspect shown on Map 3.1 that we temporarily pass over here, Crimean-Congo hemorrhagic fever, will be discussed later in Part III within our analysis of Dr. Sabine Huebner's "Balkans origin" theory of the Plague of Cyprian.

A final pathogenic suspect not covered by Dr. Kyle Harper in *The Fate of Rome* or his other scholarship on the subject but later informally suggested by him (to our team) as a differential alternative to members of the African *filoviridae* is louse-borne relapsing fever (LBRF).

LBRF is an intriguing pathogenic suspect to be sure, and in certain ways, it seems a good fit for the Plague of Cyprian. It isn't a filovirus but rather a spirochaete bacterium (*borrelia recurrentis*). Its emergence is verifiably and applicably ancient. It is biogeographically included through centering in Ethiopia and the Horn of Africa more generally in the twenty-first century, but it also formerly caused widespread disastrous epidemics across much of Europe, Asia, Africa, and even North America—until the early twentieth century. It is transmitted via the human body louse (and perhaps the head louse). Together with the closely related bacterium (*borrelia duttoni*) that causes the (not applicable) tick-borne relapsing fever (TBRF), LBRF is often considered one of two ecotypes of a single genome species. It has been, in the late twentieth and twenty-first centuries, associated with wars and famines in Ethiopia, Eritrea, and Somalia in particular. Refugee and homeless populations in Europe, particularly during an outbreak in Marseille in 2002, have also been affected. More specifically, the disease is associated with "human disasters created by war, forced migration, poverty, famine, breakdowns of personal hygiene and seasonal spikes of cold wet weather" that promote crowding and increase the risk of disease transmission. The disease it causes is highly pathogenic. Left untreated (modern antibiotics cut the CFR to under 10% in most instances), it can be between 10 to 40% or even 10 to 70% fatal. Its "most distinctive feature" is the relapse phenomenon where victims seem to recover and then relapse. Other symptoms include myalgia and rigor, profuse bleeding from the nose, and (in around 70% of patients) acute jaundice [5, 38].

Our attitude toward LBRF as a possible Plague of Cyprian pathogenic suspect, however, despite its intriguing fit in certain ways, is largely negative. The associations with social breakdown, violence, and displacement fit strongly with what happened in Alexandria in 249 CE and also with some of the later Third-Century Crisis plague outbreaks (particularly during the Gallienic period) in the early 260s CE. The biogeography, seasonality, lethality, molecular-phylogenetic emergence timeline, and climate factors all work. Taken, somewhat artificially out of context, we also can't rule out LRBF for some of those later Plague of Cyprian outbreaks after 253 CE, as the chaos of the Third-Century Crisis deepened and metastasized in large parts of the Roman Empire. The Gallienic accounts, for example, are simply too brief and vague in terms of symptomatology (we lack another Cyprian for that later decade), but the overall association with violence, presumably unhygienic conditions, refugee displacement and violence (the chief enablers of LBRF) are very strong. There is also a certain overlap in symptoms with what is described with the Plague of Cyprian between 249 and 252 CE. However, there are also reasons, in a more holistic context, to very strongly doubt LBRF for that earlier phase of the outbreak and thus, inferentially, for the whole temporal spectrum of mid-third-century pestilence. Firstly, LBRF's relationship with violence and disruption in modern times is hardly exclusive. Similar associations, as noted, have been made with deadly outbreaks of our other remaining viable candidates—the members of the African *filoviridae* (Ebola and Marburg).

Additionally, some aspects of the functioning of Roman imperial urban centers also seem to make the spread of a disease like LBRF highly epidemiologically unlikely—at least before the peak of the Third-Century Crisis disrupted a number of them in the 260s CE. Even at that peak point, as recent scholarship on the Third-Century Crisis has rightly pointed out, violence, disruption, and destruction within the empire were far from universal. This same line of revisionist reasoning applies even more strongly before the early 260s CE. Neither Carthage, Rome, nor Athens (all hit hard by the Plague of Cyprian in 251–253 CE) were in the same disrupted state at that time that Alexandria had been in two years earlier or other cities in the Roman Empire were later on, as the Third-Century Crisis peaked in the 260s. Thus, attempting to attribute these earlier outbreaks to LBRF seems to impermissibly put the cart before the horse in an analytical sense. The Roman baths, in all their varied aspects, in these hard-hit cities seemingly fully functioned in the 250s or even in the early 260s CE. Wouldn't that have dramatically limited the spread of LBRF? In a similar and supportive vein, the aged urine used to clean the clothes of the better-off urbanites at the fullers would have also strongly served as a prophylactic and likely caused the pestilence to be more discriminating along class lines. Ammonia kills lice. The Plague of Cyprian, as noted by Harper based on the contemporary evidence, was unique, indiscriminate, novel, and unknown in presentation—utterly mysterious and terrifying. LBRF was, by contrast, well known by the *corpus* of Greco-Roman medicine. Hippocrates himself had chronicled the disease as early as the fifth-century BCE on the North African island of Thasos—particularly its most notable pathology:

> The great majority (of sufferers) had a crisis on the sixth day, with an intermission of six days followed by a crisis on the fifth day after the relapse. [38]

It also is important to note what is *not* in the detailed accounts of Dionysius or Cyprian. The relapse phenomenon is not mentioned. Additionally, an apparent sixth-century outbreak of this disease was labeled "The Yellow Plague" because of the resulting epidemic jaundice. Such symptoms are not mentioned in any ancient source for the Plague of Cyprian. LBRF also does not manifest diarrhea [38]. Thus, the whole timeline and pathology of the Plague of Cyprian seems an overall very unconvincing fit with LBRF. While it is impossible for us, for example, to absolutely rule out LBRF in the last major Plague of Cyprian outbreak in certain disrupted violence-afflicted regions of the empire in 262 CE (during the reign of Gallienus) due to the paucity of specific/descriptive information regarding pathology, it seems far less logical to postulate a second new plague as opposed to a flare-up of an existing one with similarly strong means of epidemiological transmission and associations with violence without any evidence to the contrary. Nor, in 262 CE, were plague outbreaks confined to such areas. There also was a massive outbreak in Rome at that time; this is an additional telling factor against LBRF. Kyle Harper also makes a detailed and more affirmative and convincing differential and circumstantial case for a retrospective diagnosis of a VHF—especially a filovirus like Ebola or Marburg. Filovirus VHFs indeed seem to exhibit a stronger evidentiary fit, in terms of both pathology and epidemiology, with the Plague of Cyprian than any other alternative. The extremely lethal qualities of these viruses, their particular modes of transmission, a very specific list of matching pathologies, the "exotic," horrific, and terrifying pathological manifestations of these diseases could easily engender the very sort of baffled and terrified reactions (including documented modern *ville morte*-type events) that are apparent from both the surviving written sources *and* the Plague of Cyprian-related archeological evidence, detailed by Kyle Harper, that emerged between 2006 and 2014 from Egypt (see below) [11, p. 137].

Overall, Harper lays a very solid foundation for a consensus shift regarding the Plague of Cyprian. *The Fate of Rome's* rendition of that case was, however, criticized by highly respected scholars such as Haldon, Elton, Huebner et al., on detailed evidentiary grounds. Given *Fate*'s vast topical breadth and intended broader audience, however, it would have been impractical for Harper to marshal, analyze and recapitulate all of his considerable published analysis and evidence on this single topic within that vast work. In a certain sense, we have done that here. We have also added to his initial case with a more contextually developed and fully evidenced connection between recurrent sudden dry to wet climate pattern events and viral hemorrhagic fever outbreaks, the holistic application of additional primary source evidence and accompanying analysis, the application of an interdisciplinary overlay of modern scientific evidence, as well as with a series of slight refinements and an expansion of Harper's overall conclusions in terms of pathology and epidemiology. We also further supplemented his highly effective and delimiting differential retrospective diagnosis methodology. We also argue that, so far in Part I, we have strongly answered critiques of Harper's scholarly conclusions on the plague's origins, seasonality and epidemiological dissemination with additional evidence and analysis of our own. This includes our detailed analysis and rebuttal regarding the key Dionysius/Eusebius account. Our analysis so far demonstrates that, as more evidence and systems-synthesis scholarship is applied, Harper's initial theory of a filovirus as the cause of the Plague of Cyprian appears even more strongly supported. In an evidentiary sense, we believe that we have already fully evidenced or strongly begun to support #s 1–21 and #s 23–30 of the points regarding the relevant pathology and epidemiology contained in our expanded and modified List 2.1—developed by our team from the foundation of Harper's original evidentiary Table 4.1 in *The Fate of Rome*. Much more supporting evidence on all these points, along with additional evidence and analysis concerning the temporarily missing point #22, will appear in Part III.

A basic plausibility standard for Harper's filovirus theory regarding the Plague of Cyprian seems, objectively, well on the way to being met. An absolutely scientifically definitive solution to the ancient puzzle of the Plague of Cyprian will, however, probably prove elusive for the foreseeable future. As referenced above, hope was raised briefly when an apparent plague pit/disposal facility for Plague of Cyprian victims was found near the city of Luxor (south and directly upstream from Alexandria on the Nile) in Egypt. Archeological work continued at the site between 2006 and 2014. The archaeological remnants of the pit and "corpse disposal facility" at the Funerary Complex of Harwa offer potentially grim testimony to the Plague of Cyprian's hellish reality and the extreme terror that it inspired. Probably in 249 CE by our calculus (and Harper's), older mummies and coffins from the site were evidently unearthed and used as fuel for bonfires to burn victims of the plague. Large kilns were also used to cover the remains of plague victims in super-heated lime. Because of the Egyptian climate and the extreme nature of the prophylactic measures taken in Antiquity, no DNA was recoverable from any of the human remains at the site. In any case, if the Plague of Cyprian pathogen was indeed a filovirus, recoverable DNA (while it might have identified a bacterial pathogen or a DNA genome type virus such as smallpox) also *would not* have helped identify an RNA VHF/filovirus. Filovirus genomic material, like that of other RNA viruses, degrades very rapidly—in minutes to hours if not carefully preserved. 1,757 years would have been far too long [11, p. 137 and 332 (note 38); 32]. A newer archaeological-paleo-genomic technique that we will discuss In Part III, however, offers some faint more general hope in this regard.

Paleo-genomic-based science has often proved invaluable in examining and revealing the pathogenic origins and details of past plagues such as the Black Death. From a systems-synthesis point of view, this DNA/RNA viral discrepancy obviously represents a cautionary factor (in an even more analytically severe way than does surviving literary source bias in ancient history) in applying solely DNA-extraction-based scientific solutions to historical pandemics *where RNA-based pathogens may be potentially indicated by the relevant evidence.* The Plague of Cyprian clearly seems to be one such case. In line with this criterion, Huebner, in her 2021 article, mentions Harper's implication of a viral hemorrhagic fever (VHF) without rejecting it and also cites Dr. Ben McDonald's implication of Ebola in his 2020 doctoral dissertation, "Climate Change and Major Plagues in the Roman Period." {This was the one source that we wished to examine during this project that we were unfortunately unable to obtain due to an ORA embargo}. She does acknowledge broader scientific "difficulties" in terms of isolating DNA from ancient skeletons. Huebner, however, seems unaware (particularly during her discussion of the, in multiple senses uncertain, "Tomb of the Emperor Hostilian" at the Viminacium Archaeological Park in modern-day Serbia and of potential future "genomic work" at the park's vast necropolis to solve "the puzzle" of the pathogen involved) of this important DNA/RNA distinction. We also note that neither her account (nor others that we reviewed) mentioned any ancient pottery shards being found at the site or sent for scientific analysis (though bones at the site were)—though the referenced *bustum* burial involved slightly raises that possibility. This is an issue that we will return to in Part III. Overall, Huebner overlooks the clearly applicable potential for retrospective diagnosis distortion if the problem is approached from a single analytical, DNA-based, molecular-phylogenetic axis. In this case, somewhat ironically, scientific as opposed to historical. By contrast, Harper, at least as of 2021, seems to have been fully cognizant of this fundamental DNA/RNA analytical difficulty [13, p. 47; 15, pp. 157 and 170 (note 111); 29].

A great modern example of the difficulties surrounding RNA pathogens in this retrospective diagnosis regard were the tremendous difficulties surrounding the rediscovery and the eventual genetic reconstruction of the H1N1 flu virus pathogen responsible for the 1918–1920 worldwide Spanish flu outbreak that killed 50 million people. Many attempts to isolate the virus from samples preserved with early twentieth-century scientific methods, from frozen bodies and from bodies buried in coffins made of lead, failed. One viable sample was finally successfully retrieved from a frozen body buried in the Alaskan permafrost. Another was obtained and analyzed from a carefully preserved 1918 specimen from the body of a US soldier who died from the Spanish flu at Fort Jackson, South Carolina. The essential implausibility, and the analytically limiting effects thereof, of recovering viable RNA viral material from the third century are quite obvious [17].

Such factors and the passage of time, however, also potentially invalidate the attempts of Harper's critics to question (because large numbers of bodies have not yet been discovered at or near the site), the conclusions of the Italian archeological team at the Luxor site upon which *The Fate of Rome* relies. The site certainly is suggestive of a high mortality event, and in conjunction with the existing written sources regarding the Plague of Cyprian, highly so. Yet, in and of itself, it is not independent confirmation of "enormous mortality" from the Plague of Cyprian. Yet, multiple eyewitness, contemporary and recapitulated, ancient accounts seem to confirm that urban mortality was extensive indeed. For instance, Dionysius… "They deserted those who began to be sick, and fled from their dearest friends. And they cast into the street those who were half dead and left the dead like refuse, unburied." The overall evidence at the Luxor site, coupled with this brief sampling of the literary evidence, *indirectly but also strongly provides interdisciplinary consilience in terms of no less than seven elements of Plague of Cyprian pathology and epidemiology in List 2.1 (above).* The evidence from Luxor, and Tiradritti's and

Harper's conclusions about it, is also (despite a certain limited analytical circularity) strongly and mutually supported by these very Plague of Cyprian accounts of the early Church sources upon which Harper chiefly relied. Eusebius' *Ecclesiastical History*, relying on Dionysius, for instance, describes, in the context of discussing both those who died in and those who initially survived the pestilence, "the dangerous task of preparing for burial the bodies of the perfected and blessed martyrs." See pathology and epidemiology element #s 11, 17, 19, 26, 28, 29–30 in List 2.1 above [11, pp. 137–144 and 332 (note 38); 23, p. 139 (7.22.6–10); 32].

Another point in favor of narrowing our suspect pool down to the two African filoviruses is demonstrated epidemic viability in a European biogeographical setting due to the 1967 German-Yugoslav Marburg outbreak that will be discussed in greater detail in Parts II and III. Harper, however, still expresses some caution about his own convincing and well-supported but obviously less than scientifically definitive analysis, "There is much that must remain uncertain about the Plague of Cyprian." His caution is objectively rooted "in our profound ignorance about the deep history of pathogens like Ebola.... But our broadening awareness of the incessant forces of emerging disease, at the frontier between human society and wild nature, suggests a place for significant disease events in the past, like the Plague of Cyprian, caused by zoonotic diseases that wreaked havoc and then retreated back into their animal hosts" [7, p. 144]. Nevertheless, despite this lingering and probably insurmountable uncertainty, we believe that we have built, and will continue to build strong systems-synthesis-based evidentiary support for a modified and greatly expanded version of Harper's original epidemiological timeline, directionality, and pathology (including probable retrospective pathogen diagnosis). We also will provide entirely new elements and evidence in recreating our holistic model of the Plague of Cyprian.

While we thus don't disagree that some level of uncertainty is likely to remain in proposed solutions such as ours, it is the goal of this work to deepen our understanding of the Plague of Cyprian by laying out a continuing differential, circumstantial, and intensely multidisciplinary and holistic case regarding the Plague of Cyprian. Using a novel methodology that strives for consilience through highly detailed overlays of ancient and modern evidence and modeling strategies, we will move next to narrowing the indicated VHF pathogenic-suspect pool of the filoviruses to the single most likely viral evolutionary lineage involved. Additionally, we will produce evidence and more detailed conclusions regarding the probable origins, epidemiological functioning, broader mechanics and near-term and long-term impacts of the Plague of Cyprian within and on the Roman Empire. Our proposed multifaceted "solutions" in this work aren't, as noted, scientifically definitive. The overall arc of the continuum of events that we are proposing ranges, in our opinion, from the clearly plausible up to the highly probable.

Our ensuing analysis is geared toward a more detailed and yet more holistic model of what happened—the probable what, how, and why of the Plague of Cyprian—a devastating and significant event that likely contributed to the destabilization of the Roman Empire in the third century and beyond. Having used Harper's, supplemented by our own, multifaceted analysis to narrow our focus to the filoviruses, a similar examination of the members of the *filoviridae,* in terms of their properties, biogeographical dissemination, and patterns of molecular-phylogenetic evolution is thus in order in Part II. We will continue our case for a further consensus shift regarding the Plague of Cyprian by looking for the single most convincing pathogenic suspect in a horrific 1,774-year-old pandemic mystery. The "what" among the members of the *filoviridae* and their evolutionary ancestors. A defensible and probable pathogen retrospective diagnosis, a key and fascinating but controversial area of our research, is thus the initial focus of Part II.

References

1. Acuna-Soto, Rodolfo, Romero, L.C., et al., 2000 "Large epidemics of hemorrhagic fevers in Mexico," Am. J Trop Med Hyg. and also "Overview" 2008 contained within "Historical, Scientific and Technological Approaches to Studying the Climate-Disease Connection" Global Climate Change and Extreme Weather Events: Understanding the Contribution to Infectious Disease Emergence: Workshop Summary; Institute of Medicine (US) Forum on Microbial Threats Washington National Academies Press) Accessed on 4/13/2020. https://doi.org/10.4269/ajtmh.2000.62.733.
2. Ajelli, M. and Merler, S., 2012 "Transmission Potential and Design of Adequate Control Measures for Marburg Hemorrhagic Fever," *PLoS One.* https://doi.org/10.1371/journal.pone.0050948.
3. Bienert, W.A., 1978 *Dionysius von Alexandrien Zur Frage des origenismus im dritten Jahrhundert* (Berlin).
4. DeVore, David, 2020 "The Only Event Mightier than Everyone's Hope': Classical Historiography and Eusebius' Plague Narrative," *Histos* 14: 1–34. https://research.ncl.ac.uk/histos/documents/2020AA01DeVoreEusebiusPlagueNarrative.pdf. Accessed on 8/15/2020.

5 ECDC, 2016 "Facts About Louse-borne Relapsing Fever." https:www.ecdc.europa.eu. Accessed on 9/30/2020.
6 "Eusebius as a Historian: 2005 The Merits and Defects of His History," *Church History-Eusebius Pamphilus*. https://biblehub.com. Accessed on 08/5/2020.
7 Green, Bernard, 2010 *Christianity in Ancient Rome: The First Three Centuries* (London), p. 133.
8 Haldon, J., Elton, H., Huebner, S., et al., 2018 Plagues, Climate Change, and the End of an Empire: A Response to Kyle Harper's *The Fate of Rome, History Compass/Wiley* Parts 1–3 [inclusive]. https://doi.org/10.1111/hic3.12508. https://doi.org/10.1111/hic3.12506. https://doi.org/10.1111/hic3.12507.
9 Han, Barbara A., Schmidt, John Paul, et al., 2016 "Undiscovered Bat Hosts of Filoviruses," *PLoS Neglected Tropical Diseases* 10(7): 1–10. https://doi.org/10.1371/journal.pntd.0004815.
10 Hanson, J.W. and Ortman, S.G. 2017 "A Systematic Method for estimating the populations of Greek and Roman Settlements," *Journal of Roman Archaeology* 30: 301–324.
11 Harper, Kyle, 2017 *The Fate of Rome: Climate, Disease & the End of an Empire* (Princeton University Press).
12 Harper, Kyle, 2015 "Pandemics and Passages to Late Antiquity: Rethinking the Plague of Cyprian c. 249–270," *Journal of Roman Archeology* 28: 223–260.
13 Harper, Kyle, 2021 *Plagues upon the Earth: Disease and the Course of Human History* (Princeton University Press).
14 Hayman, David T.S., 2015 "Biannual Birth Pulses Allow Filoviruses to Persist in Bat Populations," *Proceedings of the Royal Society B: Biological Sciences* 282(1803): 20142591. https://doi.org/10.1098/rspb.2014.2591.
15 Huebner, Sabine, 2021 "The 'Plague of Cyprian': A Revised View of the Origin and Spread of a 3rd-c. CE Pandemic." Published online by Cambridge University *Journal of Roman Archaeology* 34(1) (June): 151–174.
16 "Jotapian," 2016 www.britishmuseum.org Accessed on 6/25/2022.
17 Jordan, Douglas, 2019 "The Deadliest Flu: The Complete Story of the Recovery and Reconstruction of the 1918 Pandemic Virus'". https://www.cdc.gov. Accessed on 6/17/2022.
18 Kelly, John Daniel, Wannier, Sarah Rae, et al., 2020 "The Impact of Different Types of Violence on Ebola Virus Transmission During the 2018–2020 Outbreak in the Democratic Republic of Congo," *Journal of Infectious Diseases* 222(12). https://doi.org/10.1093/infdis/jiaa163. Accessed on 3/18/2021.

19 Lash, R. Ryan, Brunsell, Nathaniel A., et al., 2008. "Spatiotemporal Environmental Triggers of Ebola and Marburg Virus Transmission," *Geocarto International* 23(6): 451–466. https://doi.org/10.1080/10106040802121010.
20 Meckler, Michael and Korner, Christian, 1999 "Philip the Arab and Rival Claimants in the later 240s" (*DIR Atlas*). https://www.roman-emperors.sites.luc.edu. Accessed on 7/4/2021.
21 Ndayimirije, Nestor, 2005 "Marburg Hemorrhagic Fever in Angola—Fighting Fear and a Lethal Pathogen," *New England Journal of Medicine*. https://doi.org/10.1056/NEJMp058115.
22 Orosius, Paulus, 1964 *The Seven Books of History Against the Pagans* Trans. Roy Deferrari (Catholic University of America Press, Washington, DC).
23 Pamphilus, Eusebius 1890 *Ecclesiastical History* Trans. A.C. McGiffert (Pantianos Classics).
24 Pigott, David M., Anoushka, I., et al., 2016. "Updates to the Zoonotic Niche Map of Ebola Virus Disease in Africa," Edited by Mark Jit. *ELife* 5(July): e16412. https://doi.org/10.7554/eLife.16412.
25 Pigott, David M., Golding, Nick, et al., 2014. "Mapping the Zoonotic Niche of Ebola Virus Disease in Africa," *ELife*. https://doi.org/10.7554/elife.04395.
26 Pinzon, Jorge, Wilson, James, et al., 2004. "Trigger Events: Enviro Climatic Coupling of Ebola Hemorrhagic Fever Outbreaks," *The American Journal of Tropical Medicine and Hygiene* 71 (December): 664–674. https://doi.org/10.4269/ajtmh.2004.71.664.
27 Reed Hranac, C., Marshall, Jonathan C., et al., 2019. "Predicting Ebola Virus Disease Risk and the Role of African Bat Birthing," *Epidemics* 29(December): 100366. https://doi.org/10.1016/j.epidem.2019.100366.
28 Schmidt, John Paul, Park, Andrew W., et al., 2017. "Spatiotemporal Fluctuations and Triggers of Ebola Virus Spillover," *Emerging Infectious Diseases* 23(3): 415–422. https://doi.org/10.3201/eid2303.160101.
29 Smith, Kiona, 2022 "Ancient DNA Points to Where the Black Death Began," *Nature, ARS Technica*. https://doi.org/10.1038/s41586-022-04800-3.
30 Smith, Payne, 2022 "Fathers of the Church from Festal Letter XXVII." https://www.catholicculture.org. Accessed on 6/22/2022.
31 Strobel, K., 1993 Das Imperium Romanum im '3. Jahrhundert'; Modell einer historischen Krise? (Stuttgart).
32 Tiradritti, F., 2014 "Of Kilns and Corpses: Theban Plague Victims," *Egyptian Archeology* 44: 15–18.
33 Tissot, Y., 1997 "Le rapt de Denys d'Alexandrie et la chronologie de ses lettres," in *Revue d'histoire et philosophie religieuses* (Paris): pp. 51–65.

34 Trompf, Garry, 2018, "A Foretaste of Eusebian Panegyricism in the Tenth 'Festal Letter; by St. Dionysius (the Great) of Alexandria." https://www.academia.edu/30902539/A_Foretaste_of_Eusebian_Panegyricism_in_the_Tenth_Festal_Letter_by_St._Dionysius_the_Great_of_Alexandia?. Accessed on 12/20/2020.
35 Truglia, Craig, 2020 "Was Eusebius Accurately Recounting Dionysius of Alexandria's Letter," *Church History/Orthodox Christian Theology*. https://www.orthodoxchristiantheology.com Accessed on 6/21/2021.
36 Tucker, C.J., Melocik, K.A., et al., 2000 "Correction: Reanalysis of the 2000 Rift Valley Hemorrhagic Fever Outbreak in Southwestern Arabia," *PLoS One*. https://doi.org/10.1371/journal.pone.edu.0248462.
37 Warrell, David, 2019 "Louse-borne Relapsing Fever (Borrelia recurrentis infection)." https://www.ncbi.nlm.nih.gov/pmc/articles/PMC6518520. Accessed on 8/3/2020.
38 Wollenberg Valero, Katharina C., et al., 2018. "Plant Phenology Supports the Multi-Emergence Hypothesis for Ebola Spillover Events," *EcoHealth* 15(3): 497–508. https://doi.org/10.1007/s10393-017-1288-z. https://earthobservatory.nasa.gov/features/MeasuringVegetation.
39 Zapata, Juan Carlos, 2014 "The Role of Platelets in the Pathogenesis of Viral Hemorrhagic Fevers," *PLoS Neglected Tropical Diseases*. https://doi.org/10.371/journal.pntd.0002858. Viewed on 6/17/2022.
40 Zosimus, 2004 *The New History*, Taylor Anderson, Ed. (Odin's Library Classics).

PART II

"The What and the How": Underlying Differential Virology, Molecular Phylogenetics, Host Species Ecology and Biogeographical Presence

"Here are dragons."—The Hunt-Lenox Globe (c. 1503–1507).

CHAPTER 4

Retrospective and Differential Pathogen Diagnosis

Retrospective pathogen identification, to at least a reasonably high degree of probability, is a key component in recreating and modeling the likely pathology and epidemiology of the Plague of Cyprian. Without that, an absolutely key, and centrally interconnected piece of the analytical puzzle is missing. This is a difficult process that is also not without controversy in terms of methodology. Thus, while we will lay out our main case on this point in this section, we will continue to detail additional evidence to support the conclusions arrived at here. Having differentially eliminated numerous pathogens, including many of the VHFs in Part I, we now turn directly to the remaining filoviruses.

Filoviruses are not living organisms but rather filamentous virion particles that are composed of a single strand of ribonucleic acid (RNA) and associated proteins, wrapped up in a lipid envelope. Twenty-first-century virologists have estimated, through the use of whole genome sequencing (WGS) techniques, that the modern taxonomic family of filoviruses most probably began to evolve long before the Plague of Cyprian, from a common ancestor, around 10,436 years ago. They have since evolved into a related family of viruses today. We have, in this section of our article, relied heavily upon and extrapolated from the work of Dr. Serena Carroll, Dr. Jonathan Towner, and other specialists at the Centers for Disease Control (CDC). Particularly informative was a 2013 (thus not, for instance, including the more recently discovered Bombali

M. Orsag et al., *Interdisciplinary Insights from the Plague of Cyprian*,
https://doi.org/10.1007/978-3-031-26094-0_4

ebolavirus) article, "Molecular Evolution of the Family *Filoviridae* Based on 97 Whole-Genome Sequences." It is important to note, however, that since 2013, alternative measuring methodologies to WGS now exist in the rapidly evolving field of molecular phylogenetics, and no method is universally accepted as scientifically superior in terms of accuracy [7, 15, 28].

Table 4.1 lists members of the *filoviridae* identified from mammalian (including sometimes, human) hosts. These are broken down into four *genera* containing ten viruses. More than half can be very quickly differentially eliminated as even remotely convincing candidates for connection (directly or through their evolutionary ancestors) to the Plague of Cyprian. Several recently discovered additional filoviruses that are not listed in Table 4.1 can also be decisively eliminated. These Piscine host filoviruses include two first isolated in the East China Sea in 2018 (Xilang virus [isolated in frogfish] and Huangjiao virus [isolated in filefish]) and three isolated from European perch in 2021 (Fiwi, Kandor, and Oberland viruses). They are eliminated by biogeography and the complete lack of any record of zoonotic transmission of fish viruses to humans; a piscine-human pathway is only established for a variety of bacterial infections such as *mycobacterium marinum* (commonly referred to as "Fish Handler's Disease") [11, 18, 21]. Biogeography also eliminates the recently discovered mammalian hosted *Měnglà* virus (China) and the Reston virus (Philippines). Additionally, neither of those viruses, nor the Spanish Lloviu virus, have been shown to be pathogenic in humans. There is also no evidence that the far more biogeographically relevant Bombali ebolavirus is pathogenic for humans. The other members of the *filoviridae* occur in Sub-Saharan Africa. While the very narrowly distributed Tai Forest ebolavirus has killed chimpanzees in West Africa, the single known human case was non-fatal. The remaining three species of Ebolavirus (Bundibugyo, Zaire, and Sudan), albeit highly pathogenic to humans, have not been found (via confirmed outbreaks or isolations) in either wildlife or humans east of Uganda.

While some historians speculate that now extinct and therefore untraceable and undocumentable families/lineages of pathogens may have caused historical disease outbreaks, there is essentially zero scientific evidence to support such claims and much to refute them. In our particular case of the POC, such speculation would seem to postulate an unlikely genetic/evolutionary contradiction: an entire lineage of viruses biologically, epidemiologically, and ecologically "successful enough" to

Table 4.1 Members of family *filoviridae* found in mammal hosts

Genus	*Species*	*Virus name (abbreviation)*	*Pathology in humans, scope of outbreaks and prevailing biogeography*
Marburgvirus	*Marburg marburgvirus*	Marburg virus (MARV)	Very Highly Pathogenic; Highly Lethal Very Widespread/ Multiple Outbreaks Endemic to West, Central, East, and Southern Africa
		Ravn virus (RAVV)	Very Highly Pathogenic; Highly Lethal Less common than MARV, which is in the same species Endemic to Central and East Africa
Cuevavirus	*Lloviu cuevavirus*	Lloviu virus (LLOV)	Not known to be Pathogenic (discovered in dead bats) No documented human cases Endemic to Eastern and Southern Europe (Spain, Hungary)
Dianlovirus	*Mengla dianlovirus*	Měnglà virus (MLAV)	Not known to be Pathogenic (newly discovered from a single bat) No documented human cases Endemic to Měnglà County, Southern China
Ebolavirus	*Bombali ebolavirus*	Bombali virus (BOMV)	Not Known to be pathogenic (newly discovered) No documented human cases Endemic to West and East Africa (Guinea and Kenya)
	Bundibugyo ebolavirus	Bundibugyo virus (BDBV)	Highly Pathogenic; Highly Lethal

(continued)

Table 4.1 (continued)

Genus	*Species*	*Virus name (abbreviation)*	*Pathology in humans, scope of outbreaks and prevailing biogeography*
			Highly Constrained Endemic to Central and East Africa (Democratic Republic of the Congo and Uganda)
	Reston ebolavirus	Reston virus (RESTV)	Apathogenic (inapparent human infection) No outbreaks Endemic to Eastern and Southeastern Asia
	Sudan ebolavirus	Sudan virus (SUDV)	Highly Pathogenic; Highly Lethal Highly Constrained Endemic to East Africa (Uganda and South Sudan)
	Tai Forest ebolavirus	Taï Forest virus (TAFV)	Modestly Pathogenic; Single case not lethal Highly Constrained Endemic to Western Africa
	Zaire ebolavirus	Ebola virus (EBOV)	Most Pathogenic of known filoviruses; Most Lethal Widespread, large outbreaks Endemic to West and Central Africa, South Africa

[11, 21]. Table by Dr. DeeAnn M. Reeder

cause a massive and far-flung ancient pestilence but which then disappear completely and without a trace. In fact, time itself and modern molecular phylogeneticists' documented ability to trace viral mutational evolution in a temporal sense, are key and informative (if not infallible) allies in our effort to delineate viral evolutionary stages (including those now wholly confined to the past) and pinpoint the pathogenic lineage most likely responsible for the Plague of Cyprian. As noted above, the Ebola viruses seem the logical place to begin our detailed examination among the filoviruses. Harper and Huebner both, while not specifically identifying Ebola as the pathogen involved in the Plague of Cyprian, seem to potentially implicate the virus (Harper more directly and Huebner through citation of the research conducted at Oxford University by Dr. Ben MacDonald). A more detailed discussion of the distribution, epidemiology, and evolutionary history of the remaining Bundibugyo, Sudan, and Zaire ebolaviruses is thus now warranted. Using WGS to estimate the age of the very narrowly geographically distributed but deadly (40% fatality rate) east African Bundibugyo ebolavirus and the widely distributed and extremely pathogenic Zaire ebolavirus (90% fatality rate), Carroll, Towner et al. produced data which also makes these two viruses highly unlikely candidates for strong evolutionary linkage to the Plague of Cyprian. Mutational evolution in both of these viruses is moderately rapid—7.06 $\times 10^{-4}$ Nucleotide Substitutions/Site/Year (NS/S/Y)—with a resulting TMRCA (time to most recent common ancestor) estimate of 1960 (calculated on the basis of 95% highest posterior density [HPD] range intervals) for the widely distributed Zaire ebolavirus. These data could thus be interpreted as indicating that both of these viruses only "emerged fairly recently." The best TMRCA estimate for the Bundibugyo ebolavirus, for instance, is only 2005 [7].

An alternative explanation is that much of the African-based Ebolavirus *genus* is indeed older than such WGS estimates indicate, but that the viruses within it went through a relatively recent destructive "genetic bottleneck." This would likely have involved a possible change in the Ebolavirus host population and/or environmental/geographic niche that left only a smaller, random, perhaps unrepresentative assortment of Ebolavirus strains. The first option of recent emergence is exclusionary for any sound causal linkage to the Plague of Cyprian and the second is quite problematic. Such an unpredictably transformative bottleneck event would muddy the analytical/evolutionary waters to the point where no firm conclusions or strong causal linkages to such a temporally distant event as the Plague of Cyprian would be possible due to the resulting limitation of the "ability to trace [the evolution of] the virus back through time." But one must also note the somewhat excluding biogeography in that Zaire ebolavirus outbreaks are largely restricted to west and central Africa (see CDC Map 4.1). There seems to be nothing placing the Plague of Cyprian's origin point in that area. Zaire ebolavirus, however, is by far the most prolific killer among the members of the *filoviridae*. It caused over 11,000 deaths in a single two-year west African epidemic that flared between late 2013 and early 2016 [7, 20].

Sudan ebolavirus is the remaining known virus in the Ebolavirus *genus*. At first glance, it appears to be a more promising candidate for

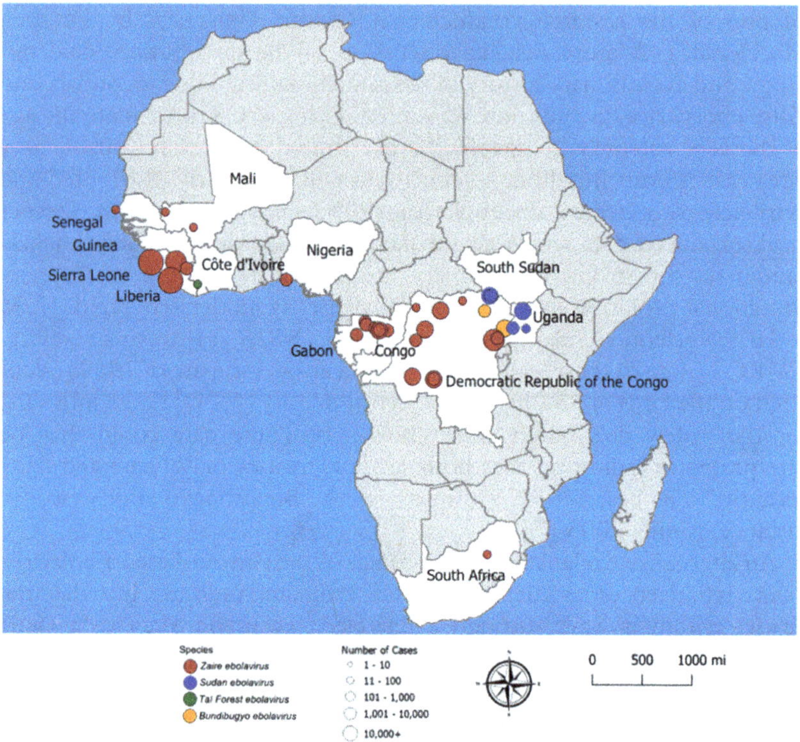

Map 4.1 Centers for Disease Control (CDC) Ebolavirus outbreaks by species, distribution, and size (Public Domain, Accessed July 30, 2022)

possible linkage to the Plague of Cyprian. It is significantly older than its viral relatives in the Ebolavirus *genus* with a much slower evolutionary mutation rate of 0.39×10^{-4} NS/S/Y; TMRCA is around 1173 CE. Thus, that common ancestor or other, more ancient Ebolaviruses from this lineage could be, in a broad temporal sense, potentially very promising candidates in context. Sudan ebolavirus' approximately 53% mortality rate and biogeography also seem to fit. While limited Sub-Saharan Roman exploration in the first century occurred, the Romans apparently very rarely visited even the northern reaches of the modern nation of South Sudan (well beyond the southern boundaries of the then Meroitic Kingdom of Nubia and its capital city of Meroe [northeast of modern Khartoum]). The outbreaks of the Sudan ebolavirus, however, are centered even farther south in what are today the nations of South Sudan, the extreme northern edge of the Democratic Republic of Congo, and Uganda. While the geography is close enough to be suggestive of the ancient sources and Harper's implication of "Ethiopia," the biogeographical gap in terms of distance would still be substantial (with operable distances somewhere around 1,000 miles) but not inherently out of the question. But a closer look at the environmental niche and pattern of modern outbreaks of Sudan ebolavirus reveals a big problem that, in another context, has been clearly noted by twenty-first-century virologists. The currently extant Sudan ebolavirus, and presumably its ancestors, appear(s) to be "highly constrained" and subject to remarkably firm "biogeographical barriers"—even in a modern world that is incomparably more mobile than the world of the third century. Consequently, a dissemination of 1,000 miles in a much earlier age without modern transportation links and less overall human ecological disturbance seems out of the question. Map 4.1: Ebolavirus Outbreaks by Species and Size effectively illustrates this pattern—with Sudan ebolavirus outbreaks essentially piled on top of one another. Sudan ebolavirus, or any similar virus from its slow-evolving evolutionary lineage, thus seem very poor candidates to have made a geographical leap across natural barriers, considerable distances, and varying biospheres to reach the distant Roman Empire, in 248/249 CE [10].

There is yet a further crucial problem, in addition to those presented by biogeography, other forms of differential analysis and broader evolutionary trajectories, with linking any of the representatives of the Ebolavirus *genus* to the outbreak of the Plague of Cyprian. The Ebolavirus *genus'* natural "reservoirs remain elusive" and are not definitively identified even by twenty-first-century virology and biology. Because this piece of the zoonotic/ecological and epidemiological puzzle is missing, our ability to tie an Ebola virus to the Plague of Cyprian via biogeographic and other ecological links to a reservoir host is nonexistent. Researchers (including one of the three main authors of this work) have devoted their lives to this quest for decades now, and the effort is very much ongoing. The reservoirs have to be there somewhere in nature, of course, but modern science is still unsure of where exactly. While certain bat species are strongly suspected and much genetic and suggestive (if uncertain) nucleic acid and serological evidence points that way, none of the Ebolaviruses have ever actually been isolated in widely suspected fruit bat or other potential host bat species "despite excellent and enormous research efforts." According to a minority of scientists, "dead end" infection of "experimentally-infected bats does not support maintenance of virus circulation in these hosts." Yet, the dominant scientific consensus is that bats are still "generally considered" to be the most likely Ebolavirus hosts. Meeting such an exacting "gold standard" of host isolation is, in fact, extraordinarily difficult and very rare for zoonotic viruses. The twenty-first-century linkages of the Marburg and Sosuga viruses to their natural reservoir host, the cavernicolous Egyptian rousette bat, are clearly the exception, not the rule. The crucial, given the grave threat it poses, hope is that Ebolavirus, soon, can also become an "exception" in this regard—through a "gold standard" scientific discovery of its natural reservoir(s) [5, 23, 35].

Simians infected with filoviruses die too fast to be included in the list of potential Ebolavirus reservoir hosts. Thus, there is no clearly defined way to envision evidentially supportable ecological paths of zoonotic transmission in the case of the current Ebolaviruses or their ancestors as possible causes of the Plague of Cyprian. As noted, apes and monkeys can carry and die from the disease, and modern Ebola outbreaks, particularly around African jungle logging sites, have been linked to contact with, butchering of, and consumption of simian "bushmeat" by workers [17, 22]. Cooking generally protects those who merely eat the meat as opposed to those who caught and prepared it. But there is no evidence that the Romans ever encountered such simian species or operated with any consistency, even tangentially or indirectly, in or near such decidedly sub-Saharan environments. The Sahara Desert itself was an imposing barrier that the Romans, despite some sub-Saharan expeditions along the Nile confined mostly to the first century, never fully overcame. Indeed, even if they had, mosquito-borne parasitic malaria infection would have been yet a further significant and deadly impediment to Roman penetration of such environments in the third century. Travelers from the then dominant Meroitic Kingdom of Nubia did make their way overland or up the Nile from the middle Nile region (around modern-day Khartoum) into Upper (southern) Egypt. The work of Dr. Solange Ashby of UCLA indeed contextualizes pilgrimages and missions to the Temple of Isis at Philae Island in the 250s and 260s CE. These were during a third-century period that she describes as the historical "height of Meroitic involvement at Philae." But contrastingly, in our systems-synthesis context, there are inherent difficulties of biogeography and ancient trade and movement patterns to consider. As noted, the southern edge of Meroitic control was still well north of the boundaries of the sub-Saharan biogeographical environment where Ebola lurks. The existing evidence is that the Romans, or even the Meroite Nubians themselves, rarely journeyed far enough south to reach such unfamiliar and inhospitable ecological regions. The arduous nature (possibly accentuated, in our context, by the drought of the 240s CE described in Part I) of such journeys also makes them highly unlikely epidemiological dissemination pathways for a filovirus. Ashby's work on the Temple of Isis at Philae furnishes an invaluable example here. Meroitic Ambassador Sasan, whose lengthy and highly temporally relevant April 10, 253 CE Demotic language inscription is recorded at Philae Island, mentioned a "long and dangerous" journey to reach Philae that probably included overland desert as well as riverine travel. He faced a similarly perilous and lengthy passage on the return trip [1, 32].

Journeys beyond the southern borders of Meroitic Nubia would clearly have been even more difficult, dangerous, against the river's flow and time-consuming. Water alone takes three months to go from the Nile's sources northward to the sea. Hypothetical ancient human journeys with highly physical portages around multiple cataracts, overland desert detours, or negotiating apparently impassable swamps, such as the Sudd (which had blocked Roman passage farther south during a first-century exploratory mission sponsored by the infamous Emperor Nero) would have either been totally impossible or exceedingly rare. Such journeys for commercial purposes would have also been completely impractical. When modern medical knowledge of infection windows and the pathology timelines related to the physical debilitation of filovirus disease are used as an overlay, we can firmly dismiss such a scenario. Ebola's comparatively rapidly evolving and highly debilitating pathology makes it particularly problematic in terms of retrospective diagnosis under such circumstances. However, "relatively rare" cases of clinical re-emergence of Ebola (weeks, months, or even years after apparent recovery) have been recently documented medically in humans and monkeys (via contact tracing in outbreaks and through laboratory macaque studies). Molecular-phylogenetic analysis in 2021 also conclusively documented such an instance in an outbreak in Guinea. Such relapse events may indicate that Ebola becomes a partially endemic disease among affected human populations; this status may, in turn, contribute to the cyclic pattern of outbreaks of the deadly virus (though this is not yet firmly epidemiologically established). Yet, this wouldn't change the overall verdict for our third-century scenario in any way. Such an inference of an Ebola relapse case as patient zero of the Plague of Cyprian would be a highly improbable "relative rarity" (perhaps no more common than 1 in 17,000 based on the two Guinea outbreaks, five years apart, that were analyzed) on top of a near and completely unsupported near impossibility in terms of ancient travel probabilities and practicalities. Ebola-harboring biogeographical environments, as identified by modern biology and medical science in the twentieth and twenty-first centuries, were, in the Roman third century, effectively inaccessible and therefore utterly alien. To put it another way, there is absolutely no record of the kitchen slaves of an imperial Roman *optimate* gastronome ever butchering a chimpanzee or of the aristocrat himself eating the simian while reclining on a *triclinium*. Further, it seems highly doubtful that a Roman *bestiarius* ever battled a Silverback Western Lowland Gorilla in the Colosseum [2, 23, 26 (Some but not all of the frightening Ebola relapse cases may have been a side effect of monoclonal antibody treatments—a phenomenon already well-documented with Argentine hemorrhagic fever), 31, 33, 34].

Any linkage of the Ebolavirus *genus*/lineage to the Plague of Cyprian thus seems, next to impossible at worst or equivalent to wholly unsupportable speculation at best. This conclusion analytically buttresses the additional difficulties relating to the retrospective diagnosis problems created by the highly uncertain situation regarding an identifiable natural host and the unclear evolutionary history of the shadowy Ebolavirus *genus*. There is more.... Individuals in the more southern part of the ranges of certain African monkey species that were known [Monkeys were kept as pets within the Roman Empire, and there is even bizarre evidence of a macaque "serving" in a fifth-century Roman legion] to the Romans from the more northern part of their ranges can die from infections of Ebola or Marburg. Yet, due to the same aforementioned interceding barriers of distance, disease, and differing biospheres, it also seems very unlikely that any Romans would have been on hand to have contact with southerly sub-Saharan populations of such contaminated animals. Such ecological/biospheric barriers are, in the twenty-first century, now codified, taxonomically, by modern biology. For example, the formerly unitary grivet monkey species (also known as the green or savannah monkey) is now generally split into multiple separate species along a north–south range axis [6, 34].

Sub-Saharan closely related species, formerly considered mere subspecies of the grivet and now known collectively as vervet monkeys, do get filovirus disease but die very fast. The sub-Saharan filoviruses are not found among their northern simian grivet kin. The far more northerly Kingdom of Aksum, located in the third century in the Ethiopian Horn of Africa and across the Red Sea into what are today the states of Yemen and Saudi Arabia, and connected strongly to Rome in terms of comparatively rapid and higher-volume seaborne commerce seems, in several regards, a far more likely Plague of Cyprian epidemiological "origin point"—linking potential epidemiology with prevailing patterns of ancient trade and movement. As noted, ancient Roman use of the term "Ethiopia" was broad and not entirely consistent. But it consistently referenced areas well beyond the boundaries of what we think of as the modern nation of Ethiopia today. Trading or other interactions, directly or through intermediaries, between Roman Egypt and North Africa and peoples farther to the south who would possibly have had more direct contact with filovirus-infected monkey populations, are remotely possible. They also seem, however, extraordinarily unlikely epidemiological transmission pathways, even when considering potential trade routes partially utilizing the Nile River, when the infection timelines and debilitating effects of highly lethal zoonotic filovirus disease are considered in toto. Furthermore, such unlikely scenarios wouldn't, in any case, favor the viruses (or their ancestors) of the Ebolavirus *genus* over those of the Marburgvirus *genus*. Ebola also seems to produce cyclic as opposed to sporadic (Marburg) outbreaks. The Plague of Cyprian seems to have been a unique event. Marburg outbreaks are concentrated in east Africa, which is also entirely on point in a biogeographical sense in terms of the implication of "Ethiopia" in multiple ancient sources [6, 37, 40, 41].

Ebola, particularly the Zaire ebolavirus species, is, however, a widely geographically disseminated killer. Ebola is generally seen as the biggest threat to humanity among the members of the *filoviridae* by twenty-first-century virological science, and there are further complications to consider in relation to it. Another complicating factor is that WGS-based molecular clock estimates (used by our team exclusively so far) are not universally accepted as definitive. Alternate measuring methodologies using a portion of the whole genome—typically either the glycoprotein (GP) sequence or, more provocatively, that portion of the filovirus genome that has been incorporated into the genomes of other animals (NERVEs: non-retroviral endogenous RNA viral elements) can produce mildly to wildly conflicting viral evolutionary timelines in comparisons with WGS estimates. Studies of Ebolavirus evolution using single GP genes have generally produced views of Ebolavirus evolution largely compatible with the work of Carroll and Towner that has informed our analysis above; they are "differing" in only "minor details" and "consistent with previous [WGS-based work]." Given that GPs, as the viral attachment protein for host cells, are under strong selection, the GP-based studies do pose additional questions of how the virus adapts in human or nonhuman primate outbreaks. The distinct possibility of a past Ebolavirus "genetic bottleneck" is also a shared element in these two differing methodologies. A fairly recent rodent-genome-based study using NERVEs, however, concluded that "Ebolavirus and Cuevaviruses have been diverging from Marburg virus since the early Miocene" (roughly 16 to 23 million years ago)! This represents a huge and potentially troubling dichotomy. The crucial yet still unsolved puzzle of the "unknown nonhuman reservoir" of the Ebola virus also further clouds the entire issue—introducing additional uncertainty, in a number of significant contexts (this one among them), to the topic [5, 37, 38, 40].

Ironically, precisely because of this extremely complicated and uncertain reality surrounding the evolutionary lineage of the Ebola virus (EBOV), one also can't conclusively rule out a hypothetical Ebola lineage ancestor or similar off-shoot, and now extinct, lineage pathogen. In either case, the entirely hypothetical pathogenic suspect possibly could share many pathological and epidemiological properties, while also possibly being somewhat different from the quite diverse Ebolavirus species of today. This suspect would lurk an unclear number of mutational-evolutionary generations back in Ebolavirus' lineage, somehow be biogeographically connected to areas that Ebolavirus species have no confirmed or strongly inferred ecological presence in and be carried by some still unknown reservoir host. That unknown host could possibly (still wholly unclear how) have been biogeographically well-positioned to interact with Aksumite and/or Roman civilization in 248–249 CE in ways that might have caused the Plague of Cyprian. Additionally, Ebolavirus or its hypothetical now-extinct ancestor would then have also somehow never detectably re-emerged in either ancient or modern times (despite the demonstrated repetitive/cyclic nature of modern Ebola outbreaks) in that biogeographical sphere. EBOV thus still seems an extremely unlikely retrospective diagnosis for the Plague of Cyprian pathogen.

Yet, further complications do arise in this regard, however, from the confusing modern overlay of human serological evidence. It indicates a potentially broader ecological niche, across nearly the entire center of Africa and beyond, for both the Ebola and Marburg, viruses. These human seropositive results include such widely spaced modern nations as Nigeria, Gabon, Ethiopia, and the broader Horn of Africa (obviously highly important in context). Methods utilized in these surveys have ranged from Immunofluorescence-based to enzyme-linked immunoassay (ELISA). Yet, some serological surveys conducted within two years of verified Marburg outbreaks in Africa found no seropositive results or Marburg seropositive results at the same levels as those in surrounding regions that had not been affected by the recent Marburg outbreak. Such results, which don't seem to have been replicated for Ebola species, have led to speculation among virologists that Marburg antibodies might fade more rapidly than those for Ebola. Due to its "double indirect" nature (measuring unique individual immune systems as opposed to isolating the virus itself) serological testing overall leads to difficult "complexities" regarding test selection, verification of validity, the significant potential for cross-reactivity, and complexities inherent in result interpretation. The effects of travel in modern settings are, for example, major wildcards in non-clinical contexts; such serological findings are thus often very far from definitive. The ongoing Covid-19 pandemic has highlighted some of these problems to the point where some experts have questioned "the continued use of point of care serological tests." Yet, other studies have shown certain types of serological tests to be, at least largely, accurate and informative when evaluated holistically. In our context, however, one applicable problem of serological testing remains absolutely central—"... the significant risk of false positives when disease prevalence is low." Regulatory pathway problems, limited availability of tests, and lack of full approval for use further complicate the picture in regard to the filoviruses. Essentially, while serological testing of a population can be useful, the likelihood of a wide range of complications (including impossible to exclude fundamental variances among individual test subjects) and the relatively high risk of false negatives and positives limit its usefulness. The resulting overlay of this "non gold standard" modern serological evidence here is intriguing but also wrought through with significant uncertainties and contradictions that limit the overlay's application and usefulness in terms of determining pathogen ecological boundaries. We have thus weighted it accordingly [3, 4, 13, 14, 16, 24, 28, 29].

Taken holistically, all this still constitutes a remotely possible yet wholly unconvincing case for Ebola (while not, as noted, *entirely* excluding it). There is, however, a far less biogeographically dubious *and* less cyclic viral suspect, with a known and clearly "included" reservoir host that lurks, in a phylogenetic sense, very close at hand to Ebola. Yet, it is also analytically notable that these two related viral evolutionary lineages, at some distant and not scientifically clearly determinable point in the remote past, were also one and the same. Both emerged from the same Origin Filovirus. It is also at least slightly possible that some other closely related but now-extinct filovirus, having emerged from the same initially joint lineage, played a role [7].

The Marburg Virus Evolutionary Lineage and the Plague of Cyprian

Figure 4.1 traces the evolution of the members of the mammalian branch of the family *filoviridae* from The Origin Filovirus on the far left (estimated at 10,436 years before the present or 8423 BCE in relation to the study publication date) branching forward in time, to the extant *genera* Marburgvirus, Cuevavirus (Lloviu), and Ebolavirus (more strains have been discovered, either through outbreaks or isolation, since then). These data were generated in 2013 by Carroll, Towner, et al. via Bayesian coalescent analysis and employing multilocus genetic sequencing [7]. As noted above, it is also at least slightly possible that a now-extinct closely related virus from the Marburg evolutionary lineage that "branched off" and then withered is our Plague of Cyprian culprit. Yet, ultimately, this possibility changes very little in our overall analysis. Using the full-family dataset of genetic sequences, Carroll et al. estimated the Marburgvirus' MRCA at 1,100 years before the present (or 907 CE in relation to the publication date). One can consider the "Origin Filovirus" (with its existence deduced by modern virology/molecular phylogenetics but not otherwise named) and the Marburgvirus' MRCA as the "grandfather" and "father" of the modern Marburg marburgvirus species. The two viruses of the Marburg marburgvirus species—Marburg virus and Ravn virus are very similar and deadly pathogens. Marburg marburgvirus virions are varied (pleomorphic)

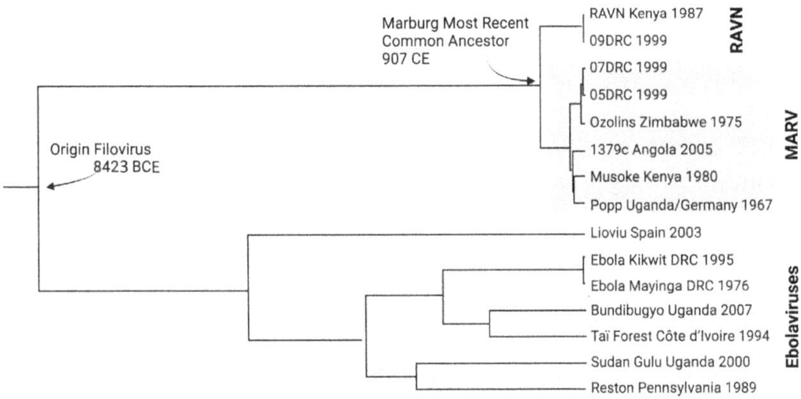

Fig. 4.1 The Marburg virus evolutionary lineage (as of 2013), with estimates for origin dates for mammalian filoviruses, including the Marburg Most Recent Common Ancestor (MRCA) (Adapted from Figure 5 in Carroll and Towner by DeeAnn M. Reeder)

particles that exhibit a rod/ring-like crook, six-shaped or branched structures. Death rates from outbreaks of these viruses range from 24 to 88% [7].

While the overall recorded official fatality rate for modern Marburg outbreaks is approximately 80%, some virologists actually suggest lower "weighted" case fatality rates for both Marburg and Ebola viruses in healthcare settings where effective palliative and other forms of basic modern medical patient care are applied. A 2016 study along these lines suggested an adjusted 65–75% death rate for the Zaire ebolavirus and an adjusted 54–61% death rate for Marburg. Such deadly but also wide-ranging figures overlap well with what ancient sources report for the Plague of Cyprian. A deduction from the aforementioned testimony of Bishop Dionysius of Alexandria, for example, produces a 62% death/de-urbanization/overall depopulation rate for the major imperial city of Alexandria during the Plague of Cyprian. As noted in Part One, deaths from multi-staged outbursts of violence as well as famine and multiple episodes of de-urbanization through flight, that all preceded or coincided with the peak stages of the plague, probably also contributed to Alexandrian mortality and exodus—an essentially multi-causal partial *ville morte* event. WGS indicates Marburg viruses to be slower evolving than other African filoviruses apart from Sudan ebolavirus. Marburg displays low genetic diversity even in its larger modern human outbreaks; this adds support for the strong likelihood that ancestral viruses in the Marburg evolutionary lineage were quite similar to the modern MARV species in pathology and epidemiology. The modest evolutionary Marburg marburgvirus mutation rate is 5.67×10^{-4} NS/S/Y. This puts the TMRCA estimate for the Marburg marburgvirus type species (effectively including RAVV strains) right around 1302 CE [7, 19, p. 14; 27].

Of the two distinct viruses within the species, Marburg is more widely disseminated and older than Ravn by a factor of around 4.2. With a slower evolutionary rate, through the Marburgvirus *genus'* MRCA, the Marburg virus is inferentially far more directly and temporally connected (with a TMRCA estimated year of 907 CE) to earlier ancestral filoviruses (and, by implication, to the Plague of Cyprian pathogen) than Ravn. Given that these two distinct viruses in the twenty-first century are still considered part of the same type species and are extremely similar in terms of their properties and effects, ancestral viruses within the Marburgvirus lineage that would have existed (as calculated by WGS-based methodology) in 249 CE, thus very likely had similar properties to both the modern type species Marburg marburgvirus MRCA and the current Marburg marburgvirus type species' "distinct" Marburg virus. Once again, it is necessary to look at the issue through the lens of biological as opposed to human-centric timescales. As noted earlier, biogeography and zoonotic ecology additionally point us very strongly in the direction of the MARV evolutionary lineage. One key caveat, however, regarding WGS-based estimates of MARV's evolutionary timeline concerns outlier extremely high-range TMRCA values for the modern Marburg virus. At the penultimate moment, our crucial allies of time, mathematics and modern molecular-phylogenetic science, seem momentarily, to desert us. The numbers appearing in our Fig. 4.1, reproduced from Carroll and Towner's article, are essentially 95% HPD interval derived/median-centered TMRCA values. Potential outlier extremely low range values (falling outside of the 95% HPD intervals utilized by Carroll and Towner) would *not* alter our identification of Marburgvirus or a Marburgvirus ancestor as the cause of the Plague of Cyprian at all; they would merely situate the subsequent stages of the evolutionary timeline for the Marburgvirus *genus* closer to the present in a temporal sense. This result would, essentially, change nothing in our overall verdict. Certain possible but not probable extremely high-range TMRCA values above the 95% HPD interval, however, would have the mathematical/temporal potential to push the emergence of the Marburg MRCA of the Marburg marburgvirus species just far enough back in time to make it, or some intermediate now-extinct closely related virus, the potential cause of the outbreak of the Plague of Cyprian in 248/249 CE [Calculations by our team on the basis of data drawn from 7]. This eventuality seems too, however, largely insignificant in terms of the continuum of our analysis and conclusions. In fact, such an outlier would only slightly strengthen our case in a certain sense by removing evolutionary steps and bringing our Plague of Cyprian suspect pathogen even closer, in an evolutionary sense, to the modern exemplar indicated by our analysis—the Marburg virus [7] (See Image 4.1).

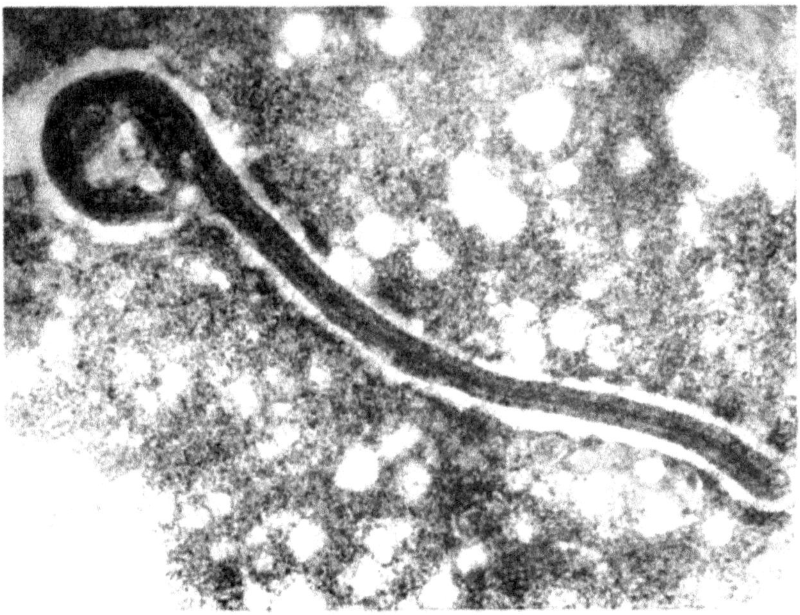

Image 4.1 The First electron micrograph of a Marburg Virion from 1967. W. Slenczka and H. D. Klenk [1967] (Image courtesy of *The Journal of Infectious Diseases* [36])

Nor is the verdict altered if one looks beyond WGS-based measurements to the much longer viral evolutionary timelines derived through the use of the alternate methodology based on NERVEs. NERVEs-based estimates would make the modern Marburg virus species a highly viable Plague of Cyprian culprit as well. As our prime suspect in the case of the Plague of Cyprian, the overall MARV evolutionary lineage thus easily best fits the entire range of scientific viral evolutionary measurement methodologies, as well as being an extremely strong fit with the pathology of the Cyprian Plague as reported in the ancient sources (an issue that we will explore again in even greater detail in Part III). After all, MARV's symptoms include myalgia, bloody diarrhea and vomiting, esophageal lesions, conjunctival injection and bleeding from the eyes, breakdown of the circulatory system leading to putrefaction of the limbs, and long-term, nerve-based damage to the senses, "sensory changes" or re-infection or persistence of the virus in the eye chamber." The disease is also, as noted, extremely lethal [25, 36]. Twenty-first-century medical pathology and deductions grounded in molecular phylogenetics thus demonstrate superb consilience with the remarkable descriptions of the third-century plague supplied by St. Cyprian and other ancient sources such as Dionysius/Eusebius. Within Africa, MARV's underlying biogeography is also broadly rooted in its confirmed presence in a known, and biogeographically "included" reservoir host species. In its sporadic outbreak patterns, MARV has also demonstrated strongly within the biogeographical setting of eastern Africa. It also possesses the necessary lethality and outbreak durability. A deadly two-year-long rural "slow burn" outbreak, beginning in 1998, in the gold mining village of Durba in the Democratic Republic of the Congo infected 154 people—killing 83% of them [12].

The outbreaks in Kitum Cave in Kenya also situate the Marburg virus closer geographically to the territory of ancient Aksum and the old Roman imperial boundaries in the Eastern African biosphere than any of the other filoviruses. A troubling but not insurmountable gap of distance (of around 1,000 miles), however, remains between this region and places like the sub-Saharan modern nations of the Democratic Republic of the Congo, Kenya, or Uganda. It would have been even further from those environments to the southeastern ancient imperial Red Sea port and border towns of Berenice, Myos Hormos, or to Luxor and Alexandria, on the Nile, in Egypt. Yet, the very wide dissemination of known Marburg virus cases is highly promising for such a leap in this context. MARV's distribution across modern-day Africa is remarkably widespread—vastly more widely disseminated than that of the other top geographical-proximity contender among the filoviruses, the highly geographically constrained Sudan ebolavirus. Sporadic and often unpredictable Marburg outbreaks have occurred in such distant locales as Uganda, Angola, Zimbabwe/South Africa, and the Democratic Republic of the Congo. Additional detection of the virus in the known reservoir host populations has occurred in nations as far away from each other as the West African nations of Gabon, Sierra Leone, and South Africa [9, 30].

Our team's conclusions above, produced in late 2020 and early 2021, were further reinforced by the world's latest far-flung, sporadic (and tragically fatal) MARV outbreak. It was revealed in August 2021 in the West African nation of Guinea. A victim died after an infection that had lasted at least nine days. Doubly confirmed post-mortem PCR testing was negative for Ebola (which would have been geographically more expected) but positive for Marburg. Medical teams from the World Health Organization (WHO), the Senegal-based ALIMA (Alliance for International Medical Action), and Guinea's Ministry of Health had been on the ground to fight a recently terminated Ebola outbreak. They fortunately rapidly contained this new and unprecedented West African MARV outbreak. Yet another unprecedented West African MARV outbreak, this time in Ghana, occurred in July 2022. Once again, this was also a region with an indigenous population of MARV's to-be-discussed reservoir host in nature—the Egyptian rousette bat [39, 41]. Marburg is a virus that can persist and disseminate, through its highly adaptable host species, across vast distances, making itself known through either outbreaks or isolations/detections, over a range of 5,800 miles. When viewed against this broader biogeographical evidentiary mosaic, the remaining geographical challenges in our context hardly seem terribly daunting [8, 11]. Our simultaneously ancient and modern, epoch-merging, intricate matrix of consilience including biogeographical factors, data generated by various branches of the applied sciences, and ancient historical evidence has been heinously difficult for us to navigate. Yet, the end result is surprisingly straightforward. Part I and II's holistic verdict, based on a multidisciplinary ancient/modern evidentiary overlap that demonstrates the ecological epidemiology and pathology reflected in the ancient historical record, the applicable biogeography, Harper's (supplemented by our own) differential analysis, as well as by application of the relevant molecular-phylogenetic evidence, the case for the MARV evolutionary lineage being the probable Plague of Cyprian pathogen seems far, far stronger than for any other disease possibility.

Two significant explanatory challenges must still be met. By what possible means was this deadly zoonotic virus transmitted to the Roman Empire's population in 249 CE given the remaining challenges of distance, terrain, and biogeography? Additionally, how did the virus do so much sustained damage to the third-century Roman Empire's population? There are obviously vastly more effective methods of medical intervention in a twentieth/twenty-first-century developing world setting than there were in the Roman third century. Yet, it is still a very considerable difficulty that a filovirus from the evolutionary lineage of MARV (which modern virologists and epidemiologists classify as a non-aerosolized pathogen having generally "low levels of contagion" and often requiring very strong and specific forms of direct contact for transmission) would have been responsible for a sustained and highly destructive empire-wide pandemic event like the Plague of Cyprian. The longest recorded modern Marburg virus outbreak was the two-year event in Durba. The deadliest was an approximately one-year outbreak during 2004–2005 in Angola that killed somewhere between 227 and 329 people. The combined fatality rate in these two outbreaks on average, however, was a terrifying 87%. While the highly lethal nature of MARV clearly fits in terms of pathology, the Plague of Cyprian was, epidemiologically, a widespread and massive pandemic…at least potentially "unlike [anything] the empire had faced before" that laid waste for at least 13 years and likely killed hundreds of thousands or more [8, 19, p. 145]. This is an explanatory burden in terms of ecology, pathology and epidemiology that we will explore through overlaying modern scientific and historical evidence. The consilience of the verdicts achieved via an overlaying analysis of ancient historical and modern scientific (medical, ecological, and molecular phylogenetic) evidence will be extended and supplemented (particularly through modeling techniques) in Parts III and IV. An explanatory metaphor for the complex, consilient, and highly interdisciplinary "systems-synthesis"-based methodological strategies that we have employed in the four parts of this work is the following:

Imagine four sheets of crystal clear plastic hung from a ceiling and fastened tightly together. On each sheet, brightly colored puzzle pieces are precisely painted. Each individual sheet's pieces contain spaces and gaps. Bright focused light behind the joined sheets projects their combined and now unified image onto a white screen. Overlapping pieces provide richer colors and details. Some missing pieces are now also fairly precisely, analytically defined by those around them and their shapes and colors can be deduced, modeled, and filled in—generating a more complete illuminated portrait of the Plague of Cyprian puzzle. Does the now more, if not yet fully complete, puzzle constitute an intelligible and realistic whole? The holistic image, in turn, must simultaneously and precisely comport with each of the component sheets and their separate sets of images. Does it? Are all the fits exact? Nothing forced awkwardly in, jumbled, or overlapping jarringly? Are there still undefined missing pieces? Even if so, does an analytically coherent, consistent, and convincing portrait still emerge? Have we also analytically eliminated any potential alternative/undefined puzzle images? Is our result, consistent, coherent, and uncluttered? Or, is it a distorted, jumbled, messy, unrealistic, an ill-fitting "funhouse mirror" outcome to be rejected? If coherent, what does this portrait reveal about the probable nature of the ancient horror of the Plague of Cyprian's likely pathology, epidemiology, ecology, and history? In undertaking such an approach, that strives for interdisciplinary and multidirectional consilience, we also overtly overlay methodologies used in the discipline of history with more experimentally based methodologies (such as modeling) rooted in the natural sciences. Thus, we hope to supportably circumvent potential problems of circularity and reductionism that might result in approaching the puzzle from a limiting single (potentially also limited by disciplinary conventions) analytical axis.

To fit the highly likely now deduced "what"—the Marburg virus evolutionary lineage—even more cohesively into the greater Plague of Cyprian analytical "whole," it is now necessary to examine the "how"—the next stage in our ancient–modern and multidisciplinary Plague of Cyprian analysis. The next chapter is thus centered on a living animal—MARV's fascinating and highly adaptable reservoir host—the Egyptian rousette bat (*Rousettus aegyptiacus*).

References

1 Ashby, Solange 2011 and 2021, "Meroitic Worship of Isis at Philae", *Egypt in Its African Context: Proceedings of the Conference Held at the Manchester Museum*, University of Manchester, 2–4 October 2009. Karen Exell, Ed. BAR International Series 2204, 2011 Archaeopress, Oxford, England and Ashby, Solange, "Rome in the Borderlands of Nubia," Slides from a Keynote Presentation at the *Association of Ancient Historians (AAH)*

Annual Meeting, University of Illinois, Urbana-Champaign [...held virtually amid the Covid-19 pandemic], May 6, 2021. All graciously shared with our team by Dr. Ashby.

2 Beeching, Nicholas, Fenech Manuel, et al. 2014 "Ebola Virus Disease," *BMJ*. https://doi.org/10.1136/bmj.g7348.

3 Borchert, Matthias, Mulanga, Subue, et al. 2006 "Serosurvey on Household Contacts of Hemorrhagic Fever Patients," *Emerging Infectious Diseases Journal*. https://wwwnc.cdc.gov/eid/article/12/3/05-0622_article. Accessed on 3/18/2022.

4 Bower, Hilary and Glynn, Judith, 2017 "A Systematic Review and Meta-Analysis of Seroprevalence Surveys of Ebolavirus Infection," *Scientific Data* 4(1): 160133. https://doi.org/10.1038/sdata.2016.133. Accessed on 12/23/2021.

5 Brown, Celeste and Quates, Caleb, et al. 2016, "New Perspectives Ebola Virus Evolution," *PLoS One*. https://journalsplos.org/plosone/article?id=10.1371/journal.pone.0160410. Accessed on 12/22/2021.

6 Butynski T.M., Kingdon, J., et al. 2013 *Mammals of Africa, Vol II, Primates* (Bloomsbury Publishing, London).

7 Carroll, Serena, Towner, Jonathan, et al. 2013 "Molecular Evolution of Viruses of the Family *Filoviridae* Based on 97 Whole-Genome Sequences," *Journal of Virology* 87(5): pp. 2608–2616.

8 CDC, 2014 "Chronology of Marburg Hemorrhagic Fever Outbreaks," Centers for Disease Control and Prevention (CDC). https://www.cdc.gov.vhf/marburg/outbreaks/chronology.html. Accessed on 6/14/2020.

9 CDC, 2018 "Deadly Marburg Virus Found in Sierra Leone Bats." www.cdc.gov/media/releases/2018/p1220-marburg-found-in-bats_html. Accessed on 8/19/2020.

10 CDC, 2020 "Ebola Virus Disease Distribution Map: Cases of Ebola Virus Disease in Africa, Since 1976," Centers for Disease Control and Prevention (CDC). www.cdc.gov.vhf/ebola/history/distribution-map.html. Accessed on 4/13/2020.

11 CDC, 2020 "Viral Hemorrhagic Fevers (VHFs)—Filoviridae," Centers for Disease Control and Prevention (CDC). https://www.cdc.gov/vhf/virus-families/filoviridae.html/. Accessed on 1/19/2021.

12 Colebunders, Robert, Tshomba, Antoine, et al. 2007 "Marburg Hemorrhagic in Durba and Watsa, Democratic Republic of Congo: Clinical Documentation, Features of Illness, and Treatment," *The Journal of Infectious Diseases*. https://ncbi.nlm.nih.gov/pubmed/17940943. Accessed on 4/15/2020.

13 Emperador, Devy, Mazzola, Laura, et al. 2018 "Diagnostics of Filovirus Detection: Impact of Recent Outbreaks on the Diagnostic Landscape,"

BMJ Global Health. https://gh.bmj.com/content/4/Suppl_2/e001112. Accessed on 4/4/2022.

14 Fierz, Walter, 2004 "Basic Problems of Serological Laboratory Diagnosis." https://pubmed.ncbi.nlm.nih.gov/14959841/. Accessed on 9/10/2021.

15 Forbes, K.M., Webala, P.W., et al. 2019, "Bombali Virus in Mops Condylurus Bat, Kenya," *Emerging Infectious Diseases* 25(5): pp. 955–957. https://doi.org/10.3201/eid2505.181666.

16 Formella, Magdalena, Gatherer, Derek, et al. 2016 "The Serology of Ebolavirus—A Wider Geographical Range, a Wider Genus of Viruses, or a Wider Range of Virulence?" *Microbiology Society*. https://www.microbiologyresearch.org/content/journal/jgv/10.1099/jgv.0.000638. Accessed on 7/12/2021.

17 Greger, Michael, "Dr. Greger Told Us About Coronavirus Risk in 2008," Video recording. *Plant-Based News* 3/25/2020. https://www.youtube.com. Accessed on 6/5/2021.

18 Harper, Kristen J. and Erickson, Kevin, 2016 "Marine Zoonotic Diseases and You," *MASNA.org*. https://masna.org/masna-education/zoonotic-diseases/. Accessed on 1/22/2021.

19 Harper, Kyle, 2017 *The Fate of Rome: Climate, Disease & the End of an Empire* (Princeton University Press).

20 Jacob, Shevin, Ian Crozier, et al. 2020, "Ebola Virus Disease," *Nature Reviews Disease Primers* 6(1): pp. 1–31. https://doi.org/10.1038/s41572-020-0147-3.

21 Kuhn, Jens H., Amerasinghe, Gaya K. et al. 2019 "ICTV Virus Taxonomy Profile: Filoviridae," *Journal of General Virology* 100: pp. 911–912 [Updated October 2020]. https://talk.ictvonline.org/ictv-reports/ictv_online_report/negative-sense-rna-viruses/w/filoviridae. Accessed on 1/21/2021

22 Kurpiers, L.A., Schulte-Herbrüggen, B., Ejotre, I., and Reeder, D.M. 2015 "Bushmeat and Emerging Infectious Diseases: Lessons from Africa," in Angelici, F.M., ed., *Problematic Wildlife. A Cross-Disciplinary Approach* (Springer, Cham), pp. 507–551. https://doi.org/10.1007/978-3-319-22246-2_24.

23 Leendertz, S.A.J., 2016 "Testing New Hypothesis Regarding Ebolavirus Reservoirs," *Viruses*. www.ncbi.nlm.nih.gov/pmc/articles/PMC4776185. Accessed on 5/16/2021.

24 Liboa, Mayara, Tavaziva, Gamuchirai, et al. 2019 "Is the Continued Use of Diagnostic Antibody Tests Justified." https://www.cochrane.org/CDO13652/INFECTN_what-diagnostic-antibody-tests-detectioninfection-covid-19-virus. Accessed on 10/14/2021.

25 Mayo Clinic, 2017 "Ebola Virus and Marburg Virus." www.mayoclinic.org-diseases-conditions/ebola-virus/symptoms-causes/syc-2035625. Accessed on 4/15/2020.
26 Mbala-Kingeni, Placide and Pratt, Catherine, et al. 2021 "Ebola Virus Transmission Initiated by Relapse of Systemic Ebola Virus Disease," *New England Journal of Medicine*. https://doi.org/10.1056/NEJMoa2024670.
27 Nyakarahuka, Luke, Kankya, Clovice, et al. 2016 "How Severe and Prevalent Are Ebola and Marburg Viruses? A Systematic Review and Meta-Analysis of the Case Fatality Rates and Seroprevalence,'" *BMC Infectious Diseases*. https://doi.org/10.1186/s12879-106-2045-6.
28 Nyakarahuka, Luke, Schafer, Ilana, et al. 2020 "A Retrospective Cohort Investigation of Seroprevalence of Marburgvirus and Ebolavirus in Two Different Ecological Zones in Uganda," https://pubmed.ncbi.nlm.nih.gov/32611400/. Accessed on 7/5/2022.
29 O'Reilly, Kevin, 2020 "SARS-CoV-2 Serology Tests: 3 Big Limitations Doctors Must Understand," *AMA*. https://www.ama-assn.org. Accessed on 3/2/2022.
30 Paweska, Janusz, T., Storm, Nadia, et al. 2020 "Shedding of Marburg Virus in Infected Egyptian rousette Bats, South Africa," *Emerging Infectious Diseases Journal—CDC* 26(12). http://hdl.handle.net/2263/80594. Accessed on 2/18/2021.
31 Pliny, *The Natural History* (5.3-8) Trans. John Bostock (London).
32 Pope, Jeremy, 2008 "The Demotic Proskynema of a Meroitic Envoy to Roman Egypt (Ph. 416)," *Echoria* 31: pp. 68–103.
33 Robitzki, Dan, 2022 "Ebola Lurking in Brain Fluid Kills Monkeys Weeks After Recovery," *The Scientist*, February. http://www.the-scientist.com. Accessed on 8/1/2022.
34 Saleh, Ibrahim, 2011 "A Roman Bestiary." www.historum.com. Accessed on 4/14/2020.
35 Schuh, Amy, Amman, Brian, Towner, Jonathan, 2017 "Filoviruses and Bats," *Microbiology Australia* 38(1): pp. 12–16. https://doi.org/10.1071/MA17005.
36 Slenczka, Werner and Klenk, Hans Dieter, 2007 "Forty Years of Marburg Virus," *The Journal of Infectious Diseases* 196(Supplement 2): pp. S131–S135.
37 Stein, R.A., 2014 "What Is Ebola," *The International Journal of Clinical Practice*. https://onlinelibrary.wiley.com/doi/full/10.1111/ijcp.12593. Accessed on 8/14/2020.
38 Taylor, Derek and Ballinger, Matthew, et al. 2014 "Evidence That Ebolaviruses and Cuevaviruses Have Been Diverging from Marburg Viruses Since the Miocene." https://peerj.com/articles/556/. Accessed on 8/14/2020.

39 WHO, 2022 "Marburg virus-Ghana," 22 July 2022. https://www.who.int. Accessed on 8/2/2022.
40 WHO, 2014 "Ebola and Marburg Virus Disease Epidemics: Preparedness, Control, Alert, Control and Evaluation." https://www.who.int. Accessed on 9/20/2021.
41 WHO, 2021 "Marburg Virus Disease-Guinea," *Disease Outbreak News—World Health Organization*. https://www.who.int. Accessed on 3/2/2022.

CHAPTER 5

Of Bats and Empires: The Egyptian Rousette Bat and the Kingdom of Aksum

The Egyptian rousette bat, one of two species in the megabat *genus Rousettus* native to Africa, has been around a lot longer than even the distant beginnings (nearly 2,800 years ago) of the Roman Empire. As a species, *Rousettus aegyptiacus* is around 2 million years old. On such a "deep time" biological scale, less than 1,800 years back to the Plague of Cyprian is an insignificant span of time. Such reckoning in terms of biological, as opposed to human-centric, timescales is thus another key component to our probable solution to the interdisciplinary puzzle of the POC. This relationship clearly traces back to the Origin Filovirus, of the MARV evolutionary lineage (at least as determined by WGS) that the bats host. Evidence for this can be found in the recently scientifically documented viral host relationships between filoviruses and the fruit bat species of the broader *genus Rousettus* (as well as an additional bat species, *Eonycteris spelaea*). These have been discovered, by multiple recent scientific findings, to display the very types of long-term biological evolutionary linkages, widespread geographical distribution, and (viral) genetic diversity embodying exactly such a "deep time" biological/ecological relationship. That ERBs coexisted with the MARV evolutionary lineage less than 2,000 years ago at the time of the outbreak of the Plague of Cyprian is thus far more than a mere analytical inference. Egyptian Rousettes indeed enjoy long-term protective immunity against Marburg. Given the lack of evidence of genetic

M. Orsag et al., *Interdisciplinary Insights from the Plague of Cyprian*,
https://doi.org/10.1007/978-3-031-26094-0_5

bottlenecks in the development of the virus and its wide dissemination within *Rousettus'* range, bat and virus have most likely coexisted smoothly over several thousand to ten thousand years. This stable relationship apparently accounts for the wide distribution of the Marburg virus across Africa. The "highly divergent viral genome sequences" within Marburg (despite MARV's slow basic mutation rate and documented record of very limited genetic divergence during human outbreaks) also confirm a very "long-term association with a suitable reservoir host of a large population size." Modern virologists and biologists have reached estimates of 5,000 (seasonally) MARV-infected Egyptian rousette bats in a single mine (Kitaka) in Uganda. The ERB is the zoonotic "primary source of human infection" for the Marburg virus. Evidence suggests that some ERBs migrate seasonally in search of food; individually marked bats have been recaptured at roosts 500 kilometers from their initial capture sites. These individual bat travel distances, combined with the extensive species range, ensures distribution of the Marburg virus over a vast area with very similar strains found in "geographically distant locations" as far apart as Gabon and South Africa. Fortunately, the inherent difficulty of zoonotic transmission and the relatively low base biological component of MARV's infectiousness makes actual human outbreaks of the deadly Marburg pathogen very sporadic over its vast range [12, 14, 16, 18, 20, 22, 30, 34, 37, 40] (Image 5.1).

The Egyptian rousette demonstrates a modern range that is "very vastly spread over a large geographic area." Even in the modern Anthropocene era of massive human ecological disturbance of the natural world, the ERB has healthy populations throughout most, if not all, of this range. Consequently, it is a species "of least concern" in an ecological (conservational) sense. Its range includes parts of North Africa, Egypt, Ethiopia, Sub-Saharan Africa, the Middle East and Cyprus, Pakistan, and northern India with genetically distinct and stable subspecies populations (though intergrades are possible) within the broader range. Of particular note in pinpointing the cause of the Plague of Cyprian, are the populations of eastern Sub-Saharan Africa (the Democratic Republic of Congo, Uganda, and Kenya), the Ethiopian highlands population, and the Nile Valley population in Egypt (as well as that of Yemen across the Red Sea) that originally gave the species its name [1, 7].

Rousettus' success is due to a combination of its behavioral adaptability and several evolutionary advantages. Egyptian rousettes, along

Image 5.1 "Stanley"—an adult male Egyptian rousette bat. While "Stanley" resides in Henry Doorly's Desert Dome, other Egyptian rousettes are housed in the Lied Jungle. This distribution at the Omaha zoo is actually reflective of the extreme habitat-adaptability and variability of this species in nature (Image Courtesy of Lindsay Sears/Amanda Knobbe—Omaha's Henry Doorly Zoo and Aquarium [Desert Dome] [Image supplied by Andrea Henning/Omaha's Henry Doorly Zoo and Aquarium])

with a handful of other species, are unique in the Old World Pteropodidae fruit bat family in possessing echolocation ability (although they use audible "tongue clicks" rather than high-frequency sound that characterizes echolocation in all other bat groups). These medium-sized nocturnal creatures are around 15 cm long and have an average wingspan of around 60 centimeters; males are larger than females. In captivity, these bats can live for twenty years or more. In nature, the average lifespan is generally around half that—largely due to chronic vitamin D deficiencies in wild populations and predation particularly (in areas where the two species ranges overlap) by the Lanner Falcon. Egyptian rousettes are remarkable in terms of the variety of habitats they can inhabit (as long as water, a suitable roost, and a supply of edible fruit is close by). These bats, distances vary seasonally, fly up to 32 kilometers to reach their food supplies [1]. Several types of deciduous forests, rainforests and jungles, mountains, savannahs, arid deserts, and semi-deserts as well as mountains are all suitable habitats.

Highly social animals, ERBs live together in colonies ranging in size from several dozen to hundreds of thousands of individuals. Their colonies are extremely noisy much of the time, as the bats chatter away to one another in colony-specific vocalization "dialects." Larger colonies however, can even amicably accommodate bats of other species. ERB colonies are often found in caves, but they can also roost in trees and a variety of human structures—mines, abandoned buildings, tunnels, tombs, granaries, ruins, etc. In areas of MARV presence, the disease is passed to conspecifics in these colonies through saliva, guano, and other bodily secretions. In larger colonies, the virus is constantly present with generally between 2.5 and 5% of the roost's population as carriers. Virus presence achieves higher peaks, however, during two periods surrounding the bi-annual (roughly February–March and August–September) birthing cycle of the species. The virus "pulses" during these bi-annual peaks, in particular, through juvenile bats, who have no prior exposure to the virus and are thus highly susceptible [14, 18, 22, 37].

At dusk, Egyptian rousette bats sally forth from the colony to find food. They aren't picky eaters—consuming ripe as well as "unripe, insect and fungus-damaged fruits." Figs seem to be central to the diet of many wild populations. Apples are a major staple of their diets among modern captive-bred populations and in the sections of their natural range in which apples are grown. Modern nutrition optimization at zoos or other settings among captive populations, and the bats' hallmark adaptability, has led to the consumption of vegetables. Cultivated crops of figs, dates, and apricots are all also regular targets, among north and east African populations, for wild bats' nocturnal feedings. Egyptian rousettes also dine on the fruits of wild carob, mulberry, sycamore, baobab, and lilac. Leaves of certain trees can also be consumed along with fruit. These bats are prodigious eaters—consuming 50–150% of their body weight in fruit during the evening. Digestion is rapid, and the bats often tend to excrete where they eat. As noted above, ERBs breed twice a year and produce one to a (rare) maximum of two pups per litter. Migration, as a general rule, is generally opportunistic and triggered by "seasonal changes in the abundance of fruit." Other migrations from the more northern parts of the species' range appear to be more climate-driven as are the altitudinal migrations of certain populations—such as that of the population of the Ethiopian highlands [4, 11, 22, 23].

The relationship of the Egyptian rousette bat to human civilization has long been complex. In the modern era, ever since the species was first described by Napoleonic-era French naturalist Étienne Geoffrey Saint-Hilaire near Giza, Egypt in 1810, these bats have continually adapted very well to human civilization. Humans indirectly furnish food and shelter. Additionally, "Egyptian rousette bats adapt well to captivity" and have long been kept as exotic pets. Indeed, a minority of captive-bred and even wild individuals can become fearless and friendly toward humans—allowing themselves to be hand fed for instance. But they have also been traditionally seen as agricultural pests. They have been hunted by humans, had their colonies attacked with poison (pesticides and insecticides), dynamite, and even sulfur. The "frugivorous nature" of the bats can, on occasion, cause extensive damage to fruit harvests. Their large, noisy, conspicuous roosts make them fairly easy targets. Their bi-annual birthing cycle, which also seems to be the necessary vehicle for biologically long-term Marburg virus persistence among the species, has been posited to be a long-established evolutionary response to "extrinsic mortality risks conferred by hunting pressure [including pressure from human hunters]." Modern research conducted around the Kitaka mine in Uganda has shown additional consequences of human activities. Eradication efforts triggered by a Marburg virus outbreak there, in fact, led to higher viral rates in surviving bats in the vicinity, raising the risk for spillover. *R. aegyptiacus* has become a perhaps unusually steady, stable, and almost predictable filovirus "reservoir" species for the deadly (for humans but not for the bats themselves) Marburg virus. The two birth pulses a year allow for continued viral presence in ERBs, especially in the larger colonies that are partially a product of the success and adaptability (particularly to humans) of the species. Many other bat species display a different pattern—where a mere single birth pulse allows for a variety of different types of viral presence to subside. The lack of a stable relationship with such a broadly adaptable successful reservoir host may, in fact, "feedback" and explain the possible genetic bottleneck and the more generally "ephemeral" nature of the Ebolavirus *genus*' ecological presence discussed earlier [3, 14, 18, 27].

Despite the risks from humans, however, these highly mobile animals are often drawn to "areas devoted to horticulture and associated with human settlements." Human civilization has also historically played a role in the greater spread of the species through the indirect generation of "anthropogenic stepping stone dispersal routes," especially in forbidding arid or semi-arid environments. The creation of artificial oases or the utilization of natural ones by humans, through such practices as planting date palms around desert cisterns, has been of "particular relevance" for *R. aegyptiacus.* These flying animals use such places to overcome landscape barriers and link otherwise potentially biogeographically isolated populations. Recent scientific surveys show a clear preference among ERBs for "human-related habitats," with 63% of surveyed individuals in one recent major study settling in such anthropogenic habitats [10].

These adaptable habits have enabled *R. aegyptiacus* to cross biogeographical barriers to reach new and altered habitats and mix with separated populations on the basis of what wildlife biologists call metapopulation dynamics. As noted above, the Egyptian rousette bat's various subspecies do sometimes mingle and intergrade over distance, creating conspecific hybridization in limited zones of overlap. Such "highly vagile and panmictic habits across large areas" also facilitate "viral transmission." These mobile bats also, however, indirectly carry out a beneficial and important role as pollinators—particularly for such plant species as the baobab tree. "As long as they [ERBs] have a place for shelter and to find food, they seem to do very well." The species is a survivor; they adapt to living conditions that would be difficult for other bat species. The Egyptian rousette bat's vast range is indeed indicative of this reality (See Map 5.1) [18, 31].

The Kingdom of Aksum and the Roman Empire in the First–Third Centuries

Romans seem to have been quite interested in bats and fairly lacking in the negative cultural associations assigned to these strange nocturnal animals in later eras. Roman naturalists reported on bat behavior and even described a particular species as "a fox with skinny wings." Given the ERB's range, abundance, and certain attraction to the Empire's intense fruit cultivation in Egypt and the Middle East (figs, dates, and apricots, among others), this description could well have been of an Egyptian rousette bat. Bats also played a healing role in Roman folk medicine and superstition. "Bat blood is said to be an antidote to snake bite, bat hearts are said to repel ants and bat organs are listed, sometimes in combination with plants, as powerful remedies against a range of illnesses" [36].

Having built extensively upon Dr. Kyle Harper's original applied evidence and differential analysis, we made an interdisciplinary case for a filovirus as the cause of the Plague of Cyprian. We have also used molecular phylogenetic and other scientific evidence to narrow the possibilities among the members of the *filoviridae* down to a single viral pathogenic lineage with a specific known reservoir host species. At this point, however, our direct evidence (particularly of the ancient historical variety), at times, becomes somewhat thinner. Our aims and methods thus will, as this chapter progresses, change subtly. We will begin to utilize the ancient–modern overlay of the evidence that we do have to develop broad ecological behavioral models and factual overlaps to support a collectively compelling and well-defined range of probabilities that acknowledge areas of uncertainty within their broader confines. To further link bat and virus to the Plague of Cyprian, we must examine the situation prevailing in the 240s CE just over the empire's southeastern border, beyond the Roman-Egyptian border town and the port city of Berenice, on the Red Sea (Image 5.2).

According to the multiple ancient Roman and later Byzantine sources utilized by Harper, the Plague of Cyprian began in "Ethiopia." We have addressed modern scholarly challenges to that verdict in Part I. To the Romans, however, this, as noted earlier, meant (more broadly) all areas in south of the Roman Empire's African borders or (more narrowly) landlocked sections of the non-Roman African interior from which the Nile originated. The narrower conception would have centered on Meroitic civilization, and the more expansive also including Aksumite civilization to the southeast and south of Roman Egypt as well as the civilization of ancient "Ghana" (another confusing appellation from a modern point of view) in what we would today call Sub-Saharan central and west Africa. The southeastern to northwestern geographical orientation vis-a-vis Roman Africa in terms of the two regionally predominant, and appropriately positioned, civilizations (Meroe and Aksum) thus potentially further validates Harper's original epidemiological dissemination pattern and timeline. Though, as noted earlier, there are also very strong reasons (virological, biogeographical, epidemiological, and related to ancient movement patterns and timelines) to strongly doubt Meroitic Nubia as a Plague of Cyprian source. The forthcoming interdisciplinary continuum of evidence that contrastingly strongly implicates

Image 5.2 "A Fox with Skinny Wings"—Adult Female Egyptian Rousette Bat and Baby in Flight (Image Courtesy of Shutterstock)

the Kingdom of Aksum, in the end, convincingly supports the epidemiological origin claims contained in ancient sources such as the *Excerpta Salmasiana II* and vice versa. As we will explore further in Part III, this includes both the narrower and broader ancient Roman conceptions of "Ethiopia." The whole ancient–modern cumulative overlay is thus mutually supporting and achieves an interdisciplinary consilience greater than the sum of its parts. After apparently originating somewhere beyond the imperial borders to the southeast and then spreading southeast to northwest into Roman Egypt, the terrible disease first hit with a destructive power perhaps most memorably illustrated by the archeological finds at the Funerary Complex of Harwa site near Luxor and Bishop Dionysius' detailed description of the lethal plague's devastating impact upon Alexandria. As noted earlier, there are three separate populations of ERBs to take note of in this context—those of Egypt's Nile Valley, those of the Ethiopian highlands in the African Horn, and those of Sub-Saharan east Africa. The Egyptian population, see Map 5.1, is of the subspecies (*R. aegyptiacus aegyptiacus*) [21, 22].

Map 5.1 Egyptian Rousette Bat distribution (Map by Dr. DeeAnn Reeder, Jack Sutton [21–22 and see also GBIF.org GBIF.org, GBIF Occurrence Downloads https://doi.org/10.15468/dl.h6uzmt; https://doi.org/10.15468/dl.szf89e; https://doi.org/10.15468/dl.zrkpgd; https://doi.org/10.15468/dl.d5j768; https://doi.org/10.15468/dl.9xg2cw, all accessed on 3 August 2022])

The Ethiopian highlands and relevant Sub-Saharan populations belong to a different Egyptian rousette subspecies (*R. aegyptiacus leachi*). The Egyptian Nile Valley population, as well as the populations of Yemen, Turkey, and Cyprus, are of either *Rousettus' aegyptiacus geoffroyi* or *arabicus* subspecies. However, these populations have never been directly or indirectly linked, in any way by modern science, to MARV. Many of the environmental, biogeographical, or genetic/taxonomic indicators of Marburg's "ecological niche" presence simply don't seem to apply to these populations. However, we can imagine another possible scenario given the undoubted ability shown by *R. aegyptiacus* in "overcoming biogeographical barriers." Sheer distance wouldn't be an obstacle. Very limited radio-tracking of individual ERBs has already revealed opportunistic movements of over 300 miles [1].

An ecologically modeled opportunistic migration southward by a portion of the Nile Valley Egyptian rousette population in reaction to the megadrought in the 240s CE and co-mingling with sub-Saharan (*R. a. leachi*) populations where MARV, or one of its ancestors, would have been endemic thus seems quite plausible given the ERB's behavioral habits and flight capabilities. Then, intergrading and a subsequent return to Egypt when drought conditions eased. After the subsequent hypothetical link to the Plague of Cyprian, the virus would possibly have later disappeared within a bat population in which it, for a variety of countervening environmental reasons, could never become truly endemic. This possibility, however, is also a poor fit with the existing ancient records regarding the origin and pattern of dissemination of the Plague of Cyprian—as well as with Harper's and our preceding analyses. Those ancient sources that describe the plague's origin say that it started outside the empire not within Roman Egypt. Additionally, despite unfavorable environmental factors, the twin birth pulses could well have hypothetically maintained the virus within the *R. a. aegyptiacus* population once MARV had entered it. Similarly, there is no reason at all to assume that the Sub-Saharan African population of *R. a. leachi* would have "counter-opportunistically" moved north out of more moist conditions *into* a megadrought in the 240s CE. Thus differentially, we are left with the Ethiopian highlands population of *R. a. leachi* as the remaining and most logical modeled source of the Plague of Cyprian. It, by contrast, fits, as we will see, very well.

Fig distribution is strongly linked to ERB presence in the Ethiopian highlands and elsewhere. These "central," for Egyptian rousette bats, fig trees and their fruit are fairly drought hardy. Eventually though, in a severe drought, and the clearly regional megadrought of the 240s CE certainly seems to have qualified, the fig tree's fruit dries up and drops off the tree. Such spoiled fruit would be of limited nutritional value and would be vulnerable to consumption by animals on the ground—it would not be a reliable source of food for the Ethiopian highlands Egyptian rousettes. Befitting their adaptability, however, Egyptian rousettes have other drought-survival strategies. Their bodies are capable of effective thermoregulation to preserve water and deal with intense heat. The Egyptian rousette bat can "effectively regulate its body temperature to adapt to reduced water availability better than other fruit bat species." The bats can also temporarily shift their diet from fruits to insects as a survival mechanism [6]. Thus, an ecological model of the Ethiopian highlands *Rousettus* population riding out the drought largely where it resided is also highly viable. Further analysis of the ancient alternatives for more exact identification of ill-defined Roman-referenced "Ethiopia" in terms of the Plague of Cyprian's origin point also provides a converging overlay with modern biology. Or to put it in another way, if not Aksumite "Ethiopia" (including the Ethiopian highlands and located to the southeast of Egypt adjacent to the Red Sea), then what? The more west and west-central African civilization of ancient "Ghana" (once again there is the potential for misleading and problematic conflation with a modern nation) would not emerge as a major trading partner for Mediterranean-based civilizations north of the Sahara until around the fifth century. Additionally, all the epidemiological disqualifiers used earlier concerning Meroitic Nubia would apply even more strongly in terms of ancient "Ghana." Long arduous and largely overland journeys undertaken by small numbers of traders across the Sahara would simply not be a viable epidemiological dissemination path for MARV [35].

Meroe, however, was, as noted earlier, a third-century Roman trading partner; Meroitic exports included gold, ivory, ostrich feathers, and "other products from the African interior." Trade routes ran up through Roman Egypt, but similar and significant epidemiological and other disqualifiers apply, nearly absolutely, in terms of a filovirus. While such considerations might fall short of entirely disqualifying a more contagious airborne pathogen (though they would probably make it unlikely), we have cumulatively presented evidence in Parts I and II (with more coming in Part III) that makes such a hypothetical aerosolized pathogen a far less likely Plague of Cyprian suspect than MARV. Additionally, much of the Meroe-Roman Egypt trade was being funneled (due to cost and speed considerations) through the African territory of Aksum and the fully or increasingly Aksumite-controlled Red Sea by the third century. Meroe was, furthermore, in decline by the mid-third century. Meroitic access to the Red Sea was increasingly dependent on an often-hostile Aksum. Aksum (along with the simultaneous attacks from a variety of Nuba "mountain" tribes) would, in fact, later destroy Meroe sometime in the early fourth century. By the mid-third century, the terminuses of such trade routes were increasingly located in Aksumite territory. Any potential fading political distinction would also have been, in our context, without a biogeographical or epidemiological difference. Similarly, modern scholarly attempts, such as those of Sabine Huebner, to dismiss the repeated ancient implications of "Ethiopian" origins regarding the Plague of Cyprian as products of a recurring Greco-Roman xenophobic *topos*, overlook the varied but consistent references to the disease's uniqueness, lack of documented recurrence, unfamiliarity, unparalleled nature and novelty in the relevant eyewitness, and/or recapitulated, ancient accounts [35]. In regards to the unparalleled and novel nature of the Plague of Cyprian see, for example... 29, pp. 315–316 (7.21); 32, p. 8 (9); 41, pp. 8–11 (1.16–21)]. Nor was this undoubtedly existing xenophobic *topos* that dated back to Thucydides used to explain every Greco-Roman pestilence; yet "Ethiopian" origin was implied repeatedly regarding the Plague of Cyprian. The plague's probable epidemiology, particularly in relation to urbanization, as determined from both the ancient evidence directly and from modeling will be explored in our Part III. This also reinforces our conclusions in terms of the Plague of Cyprian's point of origin. The interdisciplinary consilience achieved with our retrospective pathogen diagnosis also fits. Defensible alternatives also seem to be entirely lacking. The comprehensive systems-synthesis congruence of the varying puzzle pieces, for us, collectively overcomes narrower potential circularity or reductionist concerns. The interdisciplinary overlay of all evidentiary and analytical roads (ancient and modern) seems to initially indicate, with inexorable consilience, a Plague of Cyprian origin point in the Kingdom of Aksum.

While the ancient–modern and interdisciplinary overlay of evidence, analysis, and inference has brought us inextricably to ancient Aksum, there is no longer just one plausible link in the analytical chain from this point in our analysis but rather three. We'll refer to these broad "guardrail ecological models" as the **metapopulation-drought scenario,** the **human disturbance scenario**, and the **anthropogenic attraction scenario**. As we'll see, the relative paucity of surviving ancient evidence from 1,775 years ago (particularly beyond the borders of the Roman Empire) simply doesn't allow for more precision in terms of inferences derived from the overlay of the limited ancient evidence and the greater volume of applicable modern scientific evidence. While none of these scenarios are inherently more likely than the other two, they add up, when considered *in toto*, to a precisely fitting next piece of the Plague of Cyprian puzzle. Together, they show several highly plausible paths as to how the Ethiopian highlands population of Egyptian rousette bats could have been positioned, in 248 or 249 CE, to allow for zoonotic transmission of MARV or one its lineage ancestors. Having explored the available evidence regarding the long-prevailing realities of bat and virus, it is now necessary to explore the particulars of the human civilization with which the virus, through its reservoir host, would have interacted in the third century—the Kingdom of Aksum (Map 5.2).

Aksumite civilization dominated what is now the modern nation of Ethiopia (and also well beyond its current boundaries) in the 240s CE. A wealthy trading partner and often an ally of Rome's, its strategic position on the Gulf of Aden at the mouth of the Red Sea allowed it to play a key role, particularly through its port city of Adulis, in both regional maritime trade and the more far-reaching annual monsoon-wind aided trade between India and the Roman Empire. Despite considerable urbanization and the existence of a written Ge'ez language, the most reliable and detailed information about Aksum usually comes from Roman sources. In a very narrow sense, the trail of breadcrumbs of evidence thus stops at the Aksumite border. By the mid-third century, however, trade and other relations between Rome and Aksum would only have grown more important given Aksum's greater degree of territorial control as well as military/territorial expansion in Ethiopia, also into what is today Sudan, and across the commercially critical Red Sea along the coastline from Aden in modern-day Yemen to Mecca in what is now Saudi Arabia. Aksum was, in the mid-third century, on the verge of its historical peak period of wealth and power. Roman appetite for spices, metals, jewels, silk, and other goods obtained from Africa, the non-Roman Middle East or (farther afield from India and China) had, apparently, remained unabated. About this trade, we know a surprising amount. An admittedly somewhat triumphalist mid-second-century oration from Sophist philosopher Aelius Aristides gives us some idea of the sheer scale of the largest maritime

Map 5.2 Ancient Aksum at its Pinnacle (Modified map [originally by Geography Skill Builders] [public domain]. The area controlled by Aksum in the mid-third century would have been somewhat less extensive)

trade, conducted through various regional centers—the great cities of Roman Africa and the Middle East such as Alexandria, Myos Hormos, Berenice, Thebes (Luxor), Jerusalem, Carthage, Antioch, and ultimately across the Mediterranean to Rome itself and the rest of the Empire:

11. For whatever is grown and made among each people cannot fail to be here at all times and in abundance. And here the merchant vessels come carrying these many products from all regions in every season and even at every equinox, so that the city appears a kind of common emporium of the world.
12. Cargoes from India and, if you will, even from Arabia the Blest one can see in such numbers as to surmise that in those lands the trees have been stripped bare and that the inhabitants of those lands, if

they need anything, must come here and beg for a share of their own. Again, one can see Babylonian garments and ornaments from the barbarian country beyond arriving in greater quantity and with more ease than if shippers... had but to enter the port of Athens. Your farms are Egypt, Sicily and the civilized part of Africa.

13. Arrivals and departures by sea never cease, so that the wonder is, not that the harbor has insufficient space for merchant vessels, but that the sea has enough, [if] it really does? [5].

Such trade was also interconnected with the central mainstay of the Roman imperial economy, the "Annona system" that supplied the roughly 920,000 inhabitants of the city of Rome itself with food from the empire's African "granary" provinces. Ancient cities with populations of 100,000 or more needed to rely on seaborne trade as it was much more efficient and economically viable and sustainable than large urban node food supply via overland routes. Evidence from the end of the third/beginning of the fourth century indicates that, in the Roman case, overland transport was 28 to 56 times as expensive as transporting the same tonnage of food via maritime means! That this basic equation applied, in the mid-third century, to the smaller cities of the Roman Middle Eastern provinces was also made anecdotally clear back in Part I via evidence from the *Tractate Taanit*'s Chapter 3 [8, pp. 12–13 and 24; 15, p. 319; 38].

Like the Roman African provinces, the Kingdom of Aksum was also food rich with a variety of grains (wheat and barley), fruits (including figs and dates), and other crops cultivated. While the ancient Ethiopians lacked the technical know-how to build aqueducts like those of the Roman Empire, they did build dams, boasted cities with populations in the approximately 10,000–30,000 range, maintained a sophisticated system of irrigation, and practiced widespread animal agriculture. Aksum incorporated portions of the Ethiopian Highlands within its borders and also developed overland/riverine trade routes that crossed over the Blue Nile (with its Ethiopian highlands origins) and connected the interior of Africa with the Roman Empire by using the Nile itself to bring ivory, exotic animals, and other goods to Roman Egypt. Rome's appetite for such commodities seemed insatiable, and Aksum's cities grew larger and more impressive. Coastal towns and settlements in Aksum furnished tortoise shells, cinnamon, gems, precious metals, gold, salt, and cloth. Gold, salt, and other types of mining were also an important part of Aksum's economy. Roman and Roman-Egyptian influence within Aksum was extensive and profound; Roman coinage, cultural trends, architecture, mining techniques, and agricultural methods were extensive in their presence and were often adopted by the Aksumites. Intensive demand for timber, metals, and other targets of mining operations and the agricultural methods practiced, as in the Roman Empire, eventually led to deforestation and soil depletion in Aksum. It is notable, in our context, that a 2017 study of recent filovirus outbreaks in West Africa found that the outbreaks were "significantly associated" with deforestation, as are many modern epidemics. Aksum's cities also featured Egyptian obelisks and Roman basilicas; Christianity later also spread from Rome to Aksum. It is quite possible that Roman colonies of traders, technical experts, and diplomats were more or less permanently stationed in Aksum. Ancient Roman sources from the late first century, such as the noted Roman poet and satirist Marcus Valerius Martialus (Martial) had also already recorded the presence of apparently not-insignificant numbers of "Ethiopian" (presumably including both Meroitic Nubian and Aksumite) people in the city of Rome itself [9, 17, 24, 25, 28].

The burgeoning role of Aksum in trade and its crucial further participation in the more far-flung trading system between India and the Roman Empire are chronicled by a remarkable Roman late first-century source, *The Periplus* (alternate spelling Periplous) *of the Erythraean Sea* (see Map 5.3). It is important to note that, by 250 CE, Aksum's continued expansion would have taken it farther than where its borders had stood in the first century—nearly to the tip of the horn of Africa incorporating such "near side" market towns as Avalites, Malao, Mundus, and Mosylon (in modern-day Somalia).

The main "Ethiopian" trading center for Roman-Aksumite commerce across the centuries was the city of Adulis, "a port established by law." It was reached by sailing from the Roman Red Sea ports of Berenice or Myos Hormos ("Mussel Harbor" in the *Periplus*). Three days' overland journey inland from the coast was the Aksumite city of Coloe—"the first market for ivory"—in the Ethiopian highlands. Five days further in was the Axumite capital city of Aksum itself (significant amounts of ivory were available there as well). Roman traders seem to have sometimes journeyed inland to obtain ivory, perhaps to cut out the Aksumite "middlemen" in the port city of Adulis. Trading season in terms of sailing conditions for departures from Roman Egypt ranged annually from January to September with the last ships returning home from "near side" or "upper market" port towns by September; probably with an eye to the looming

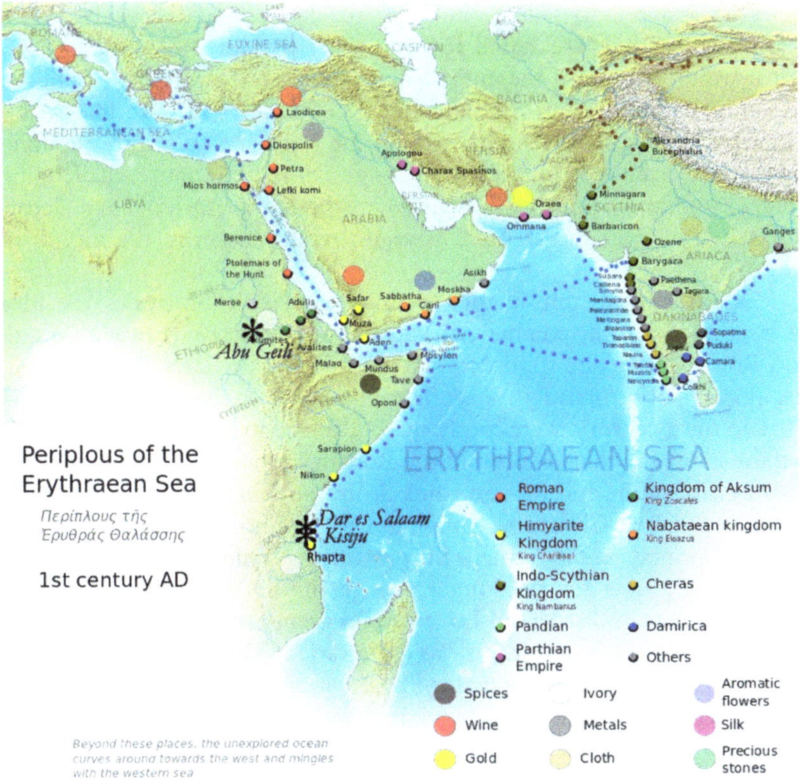

Map 5.3 The Periplous of the Erythraean Sea {first century CE} (Modified from the Map by Dr. Wilfred Harvey Schoff, 1911 [public domain] [33]. As noted, Aksum's territory would have expanded by the third century)

mare clausum. The, at times, in the eyes of the mix of Roman, Egyptian, and Greek traders, "barbarous" natives [conditions would almost certainly have improved by 248/249 CE with the greater military reach and political integration achieved by Aksum] offered ivory, palm oil, tortoise shells, frankincense, myrrh, other types of incense, cinnamon, gums, and spices (sometimes obtained from India). Voyages from there featured a usual start date of September. Traders from the Roman Empire offered cloth, coinage, glass, wheat, copper cooking utensils, weapons, wine, and olive oil. The *Periplus* offers only maritime distance values. Overland travel times as opposed to sailing times (as the latter would have been too seasonally variable) are stated for the upper part of this trading route, but it would seem as if the entire sailing route to the furthest reaches of Aksum and the Red Sea Straits could be covered (one way) in about a month. The "far side" markets around the horn and to the south along the east African coastline featured longer times and more challenging sailing conditions, which apparently necessitated a more defined departure window from Egypt. Such voyages usually began in the month of July [33].

Modern-derived numismatic evidence specifically buttresses the idea of continued strong commercial and cultural contacts between Aksum and Rome during the third century (and during the later Roman imperial period more broadly). Aksumite coins, which began to appear in the late third century, were modeled on Roman coinage—"including the weight standard which existed in the Roman Empire." A large combined hoard of Aksumite and Roman coins (from the fourth century with some displaying shared Christian symbology) was found on former Aksumite territory in modern-day Yemen. But, as noted, Aksum's territory was also a key staging area for the longer voyages undertaken by Romans and other traders to India. Varied Indian spices accounted for the bulk of that more far-flung trade, but it too was quite varied. Those voyages were also begun from Egypt in July and aided by the monsoon winds. Traders from the Roman Empire sent beryl, coral, high-value slaves, high-quality Italian wines and tin, topaz, glass, weapons, and gold and silver coins. In the first century, the Roman Empire's currency was more valuable than any of the various Indian states' currencies (though this may well have changed by the mid-third century with ongoing imperial monetary devaluation). Roman traders returned home with agate, carnelian, muslins, mallow cloth, pearls, ivory, silk, pepper and other spices, diamonds, sapphires, and tortoise shells. Some first-century Roman traders even made it as far as the mouth of the Ganges. There, they obtained "the finest" Gangetic muslin, excellent pearls, high-quality gold, and the "best tortoise shells." By the mid-third century, it is unclear whether the traders making the monsoon-wind-aided passage to India (and also obtaining access to silk and other goods from China) would have been largely Roman, Aksumite, Persian, or some mixture. Roman influence was beginning to decline, the earlier Antonine Plague had disrupted trade, and as the Third-Century Crisis progressed, direct Roman control, organization, and participation with the Indian trade gradually (and then later precipitously) declined only to revive in the fourth century. The Plague of Cyprian was, in fact, a likely further contributing factor in this regard. Aksum, in the meantime, gained full or nearly full control over the Red Sea Straits from Meroe. Persian interference and influence in the area had also grown [2, 26, 33, 39].

But the Roman-Indian trade link clearly wasn't ever wholly severed (or Roman prices for and availability of the highly prized Eastern goods significantly affected) until much later—indeed until well after the traditional date for the Fall of the Western Roman Empire in the late fifth century. Such notable changes in the parameters of the trade (route closures and/or dramatically rising prices) only occurred during the sixth century. Once again, an overlay of modern numismatic finds of Roman coins (including a number of specifically third-century coins) from India, China and even from places as far afield as Thailand, Indonesia, and Vietnam demonstrate the potentially truly vast nature of these ancient trading networks—with their principle maritime components all passing through the Aksumite-dominated Red Sea. Some caution might, however, be in order regarding the latter more exotic non-Sino-Indian numismatic finds as some of those are "lacking in archaeologically secure context." Nevertheless, third-century Aksum was clearly and simultaneously a major trading destination and trading partner for Rome, a key maritime way station, a commercial crossroads, a maritime linchpin and hub on the way to and from the Roman Empire to the fabled East. Aksum was also, tellingly in our context, most specifically and directly linked to the Roman Empire through Roman Egypt to the northwest [13, 19].

References

1 African Chiroptera Report (ACR), 2021, V. Van Cakenberghe, Victor and Seamark, Ernest C.J. eds., *Rousettus aegyptiacus* account, African Bats NPC, Pretoria. https://africanbats.org/publication/african-chiroptera-report-2021/. Accessed on 7/22/2022.

2 "Aksumite Coinage," 2018 *Khan Academy*. https://www.khanacademy.org/humanities.art-africa/east-africa2ethiopia/a/aksumite-coins. Accessed on 3/28/2021.

3 Amman, B.R., Nyakarakha, L., et al. 2014 "Marburgvirus Resurgence in Kitaka Mine Bat Population after Extermination Attempts, Uganda," *Emerging Infectious Diseases* 20(10). https://doi.org/10.3201/eid2010.140696.

4 Amman, B.R., et al. 2020 "Isolation of Angola-like Marburg Virus from Egyptian Rousette Bats from West Africa," *Nature Communications*. https://doi.org/10.1038/s41467-020-14327-8.

5 Aelius Aristides, 2003 "The Golden Age: An Oration to Rome c. AD 150," in Lewis, Jon, ed., *The Mammoth Book of Eyewitness Rome* (Carroll and Graf), pp. 339–346.
6 Armoogum, Nikhail, Schoemann, M. Carrie, et al. 2019 "The Relative Influence of Abiotic and Biotic Factors on Suitable Habitat of Old-World Fruit Bats Under Current and Future Climate Scenarios," *Mammalian Biology*. https://doi.org/10.1016/j.mambio.2019.09.006.
7 Bachorec, Erik and Horachek, Ivan, et al. 2020 "Spatial Networks Differ When Food Supply Changes: Foraging Strategy of Egyptian Fruit Bats" *PLoS One* 15(2): e0229110. https://doi.org/10.1371/journal.pone.0229110.
8 Bardi, Ugo, 2018 *The Seneca Effect* (Springer, Cham, Switzerland).
9 Butzer, Karl W., 1981 "The Rise and Fall of Axum, Ethiopia: A Geo-Archaeological Interpretation," *American Antiquity* (Cambridge University Press) 46(3): pp. 471–495.
10 Centeno-Cuadros, A. and Hulva, P., 2017 "Habitat Use But Not Gene Flow Is Influenced in Two Ecotypes of Egyptian Fruit Bat (*Rousettus aegyptiacus*)," *Molecular Ecology*. https://doi.org/10.1111/mec.14365.
11 "Diet Card Keeper," Henry Doorly Zoo and Aquarium—Bat Mix," [Internal Document].
12 Dovih, Pilot, et al. 2019 "Filovirus-Reactive Antibodies in Humans and Bats in Northeast India Imply Zoonotic Spillover," *PLOS Neglected Tropical Diseases*. https://doi.org/10.1371/journal.pntd.0007733.
13 East, Gordon W., 1938 *The Geography Behind History* (W.W. Norton), pp. 174–175.
14 Han, Barbara, et al., 2016 "Undiscovered Bat Hosts of Filoviruses," *PLOS Neglected Tropical Diseases*. https://doi.org/10.1371/journal.pntd.0004815.
15 Hanson, J.W. and Ortman, S.G., 2017 "A Systematic Method for Estimating the Populations of Greek and Roman Settlements," *Journal of Roman Archaeology* 30: pp. 301–324.
16 Happold, Meredith, 2013 "Rousettus aegyptiacus," in *Mammals of Africa, Vol IV* (London), pp. 289–292.
17 Harkness, Hannah and Tefera, Tefera, M., 2004 "Agriculture in Aksum" and "Recent Evidence of Animal Exploitation in the Axumite Epoch, 1st–5th Centuries," *Tropical Animal Health and Production*. https://doi.org/10.1023/B:TROP.0000012100.96577.5f. Accessed on 4/27/2020.
18 Hayman, David, 2014 "Biannual Birth Pulses Allow Filoviruses to Persist in Bat Populations," *Proceedings B*. https://doi.org/10.1098/rspb.2014.2591.
19 Hoppal, Krisztina, Adhityata, Shinatria, Hui, et al. 2018 "All That Glitters Is Not Roman, Roman Coins Discovered in East Java, Indonesia: A Study on New Data with an Overview on Other Coins Discovered Beyond India,"

Public Knowledge Project (Budapest). https://doi.org/10.17204/dissarvh.2018.461. Accessed on 3/31/2021.
20 Kajihara, M., Hang'ombe B.M., et al. 2019 "Marburgvirus in Egyptian Fruit Bats, Zambia," *Emerging Infectious Diseases*. https://doi.org/10.3201/eid2508.190268.
21 Korine, C., 2016. *Rousettus aegyptiacus*. The IUCN Red List of Threatened Species 2016: e.T29730A22043105. https://dx.doi.org/10.2305/IUCN.UK.2016-2.RLTS.T29730A22043105.en.
22 Kwiecinski, Gary and Griffiths, Thomas, 1999 "Rousettus aegyptiacus," *Mammalian Species* 611: pp. 1–9.
23 Lincoln Park Zoo, 2018 "Egyptian Fruit Bat." www.lpzoo.org/animal/egyptian-fruit-bat. Accessed on 4/25/2020.
24 Marcus Harold, 1994 *A History of Ethiopia* (University of California Press, Berkeley).
25 Martial, 2003 "Immigrants in Rome c. AD 90," in Lewis, Jon, ed., *The Mammoth Book of Eyewitness Rome* (Carroll and Graf), p. 285.
26 Munro-Hay, Stuart, 1989 "The al-Madhariba Hoard of Gold and Late Roman Coins," *The Numismatic Chronicle* 149: pp. 83–100.
27 Olejnik, Judith, Muhlberger, et al. 2019 "Recent Advances in Marburgvirus Research," *F1000 Research*. https://f1000research.com/articles/8-704. Accessed on 9/8/2020.
28 Olivero, Jesus, John, E., et al. 2017 "Recent Loss of Closed Forests Associated with Ebola Virus Disease Outbreaks," *Scientific Reports/Nature Research*. https://doi.org/10.1038/s41598-017-14727-9.
29 Paulus Orosius, 1964 *The Seven Books of History Against the Pagans* Trans. Roy Deferrari. (Catholic University of America Press, Washington).
30 Paweska, Janusz, T., Storm, Nadia, et al. 2020 "Shedding of Marburg Virus in Infected Egyptian Rousette Bats, South Africa," *Emerging Infectious Diseases Journal-CDC* 26(12). https://doi.org/10.3201/eid2612.202108.
31 Pigott, David, Golding, Nick, et al. 2015 "Mapping the Zoonotic Niche of Marburg Virus Disease in Africa," in *Transactions of the Royal Society of Tropical Medicine and Hygiene*, 3/27/2015. https://doi.org/10.1093/trstmh/trv024.
32 Pontius the Deacon, 2013 "The Life and Passion of Cyprian," *The Complete Works of Saint Cyprian of Carthage* Translation Editing: Nicene Fathers [1885] Philip Campbell, Ed. Introduction by Ryan Grant (Evolution Publishing), pp. 3–18.
33 Schoff, Wilfred, trans. 1912 *The Periplus of the Erythraean Sea; Travel and Trade in the Indian Ocean*, "The Voyage Around the Erythraean Sea," pp. 22–23 (1–8). https://archive.org/stream/cu31924030139236/. Accessed on 4/20/2020.

34 Schuh, Amy, Amman, B.R., et al. 2017 "Egyptian Rousette Bats Maintain Long-Term Immunity Against Marburg Virus Despite Diminished Antibody Levels," *Scientific Reports*. https://doi.org/10.1038/s41598-017-07824-2.
35 Sofferan, Steve and Wood, Sarah, 2002 "Ancient Meroe" and "Ancient Africa" Purdue University, Lectures 29–30. https://web.ics.purdue.edu/~rauhn/ancient_africa.htm. Accessed on 3/31/2021.
36 Tommasi-Moreschini, Chiarra O. 2011, "Bats in Greco-Roman Antiquity," *BATS Magazine: Media and Education* 29(2). https://batcon.org/article/bats-in-greco-roman-antiquity/. Accessed on 7/20/2020.
37 Towner, Jonathan, et al. 2009 "Isolation of Genetically-Diverse Marburg Viruses from Egyptian Fruit Bats," *PLoS Pathogen*. https://doi.org/10.1371/journal.ppat.1000536.
38 *Tractate Taanit* Ch.3: "Regulations Concerning Occurrences on Account of Which Fast-Days Are Stopped, or Alarms Are Sounded. When Fasting on Account of Rain Is Stopped," pp. 48–52, *Jewish Virtual Library*. https://www.jewishvirtuallibrary.org. Accessed on 4/14/2020.
39 Williams, J. 1997 *Money: A History* (The British Museum Press, London).
40 Xing-Lou Yang, Yun-zhi Zhang, et al. 2017 "Genetically-Diverse Filoviruses in *Rousettus* and *Eonycteris* spp. Bats, China, 2009 and 2015," *Emerging Infectious Diseases*, March. https://doi.org/10.3201/eid2303.161119.
41 Zosimus, 2004 *The New History*, Taylor Anderson, Ed. (Odin's Library Classics).

CHAPTER 6

Modeling an Ancient Zoonotic Outbreak

In light of our (based initially upon Harper's) ecological *and* epidemiological conclusions regarding the origins and dissemination of the Plague of Cyprian, this additional historical (in terms of ancient patterns of maritime commerce) analytical convergence on Aksum is, in multiple ways, a strong example of consilience. But once again, our (admittedly "guardrail") modeling must coherently "fit" with the complex overlay of *both* ancient historical evidence *and* modern science. Does it? It is within this broader analytical milieu that the three aforementioned scenarios regarding potential positioning of bat and virus in Aksumite Ethiopia in the 240s CE for the outbreak of the Plague of Cyprian must be considered. The **metapopulation-drought scenario** incorporates known behaviors of the Egyptian rousette bat to model an opportunistic migration south on the part of the Ethiopian highland's population in response to the severe drought of the 240s CE that, in this scenario, overwhelmed the species' drought-coping skills. An allegorical passage from Cyprian's work "An Address to Demetrianus" is apt here. Describing a drought-stricken "decaying world," Cyprian comments, "The fig tree says He [the Holy Spirit] shall not bear fruit, and there shall be no blossom in the vines." Such a situation could indeed have caused the Ethiopian highlands' *aegyptiacus leachi* population to depart, given the primary place of figs in the Egyptian rousette bat's diet. Then, an intermingling with the known MARV-carrying Sub-Saharan east African population and a

M. Orsag et al., *Interdisciplinary Insights from the Plague of Cyprian*,
https://doi.org/10.1007/978-3-031-26094-0_6

subsequent return to the Ethiopian highlands. This scenario would not even require any intergrading as the two populations belong to the same subspecies. This is a highly plausible scenario due to the known biological role of Egyptian rousette bat metapopulation dynamics "in filovirus ecology." Additionally, genetic connections between MARV strains isolated in Egyptian rousette bats in "geographically distant locations" clearly rule in the strong possibility of just this sort of event in light of the species proven ability in terms of "overcoming biogeographical barriers and allowing for virus circulation over broad geographic ranges" [3, 11, pp. 102–103 (20)].

The anthropogenic attraction scenario models a shorter-range foray (an extension of the Ethiopian highlands population's annual altitudinal migration) toward cultivated fruit supplies in Aksum in response to the drought or, alternatively, when it suddenly broke. As noted earlier, ERBs have been proven to be attracted to such anthropogenic environments. The irrigated agriculture of third-century Aksum would have served as a last or first available water source and cultivated fruit crops would be associated with it. Anthropogenic roost environments, such as mines, would also be viable attractors. This, however, would require MARV to be endemic within the Ethiopian highland's population of *Rousettus aegyptiacus*. This is a critical question of modern evidentiary overlay, in our context, with a very highly suggestive but not yet fully confirmed answer. Isolate testing for the presence of the Marburg virus in the Ethiopian highland's population of the Ethiopian rousette bat is yet to be carried out as there have been no known outbreaks of the highly "sporadic" Marburg disease in modern Ethiopia. A scientific study by Dr. Nick Golding of the University of Melbourne and Dr. David Pigott of the University of Washington, *extremely notably in this context*, suggests that the Marburg virus is strongly present in the Ethiopian highlands *Rousettus* population, and that modern Ethiopia has essentially been very lucky in regards to potential Marburg outbreaks (their findings are summarized in Graph/Map 6.1). Using the sort of satellite data referenced in Part I, zoological studies of *R. aegyptiacus* populations, environmental surveys geared to finding certain types of Marburg-enabling environments (such as Karst topographical landscapes featuring soluble rock sinkholes and caves), Pigott and Golding utilized mathematical regression tree modeling to calculate "relative comparison of risk across Africa" in order to head off "the potential risk of more widespread outbreaks [of the Marburg virus]" both within known outbreak areas and "Set 2" countries—those found to be at risk but without a history of modern outbreaks. For some of these "Set 2" at-risk countries, evidence of positive serology (the at times contradictory and less than "gold standard" nature of such serological evidence was, however, discussed earlier in Part II) has been found, which Pigott and Golding noted as potentially suggestive of "past [MARV] exposure in humans." Although, to date, no isolate-based studies have attempted to survey for MARV in Ethiopian bats or humans (and thus no evidence has been found). Pigott and Golding's research and analytical inferences received particularly stunning further validation in terms of the unprecedented fatal 2021–2022 MARV outbreak in faraway Guinea in West Africa (their modeling had correctly predicted MARV presence there). Researchers at the World Health Organization (WHO) have come tentatively at least, to similar even broader conclusions, "… the geographic distribution of Marburg virus may [generally] overlap with the range of *Rousettus* bats." In July 2022, both the WHO's suspicions and the modeling of Pigott and Golding received yet further confirmation with a second unprecedented fatal West African Marburg outbreak in Ghana [6, 25, 35–37].

Graph/Map 6.1 reproduced here from Pigott and Golding "Mapping…" Predicted geographical distribution of the zoonotic niche for Marburg viruses using model 1—human index cases only. Panel A shows the total populations living in areas of risk of zoonotic transmission for each at-risk country. The gray rectangle highlights countries in which index cases of disease have been reported (set 1); the remainder are countries in which risk of zoonotic transmission is predicted, but in which index cases of Marburg virus disease have not been reported and have more than 100 at-risk pixels (set 2). These countries are ranked by population-at-risk within each set. The population-at-risk figure in 100,000s is given above each bar. Panel B shows the predicted distribution of zoonotic Marburg viruses. The scale reflects the relative probability that zoonotic transmission of Marburg viruses could occur at these locations; areas closer to 1 (red) are more likely to harbor zoonotic transmission than those closer to 0 (blue). Countries with borders outlined are those which are predicted to contain at-risk areas for zoonotic transmission based on a thresholding approach. The area under the curve statistic, calculated under a stringent cross-validation procedure is 0.64 ± 0.12. Solid lines represent set 1 while dashed lines delimit set 2. Areas covered by major lakes have been masked white [25]. Image courtesy of

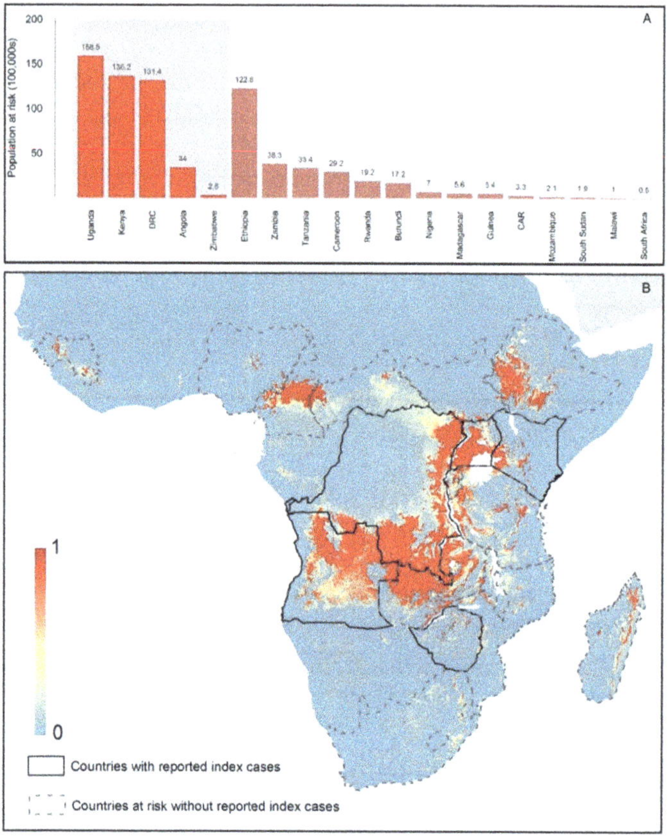

Graph/Map 6.1 MARV's Zoonotic Niche

Dr. David Pigott and *Transactions of the Royal Society of Tropical Medicine and Hygiene*

Pigott and Golding's calculations indeed revealed that "a much broader region is at risk of zoonotic transmission" of MARV than the countries (such as Zimbabwe, Kenya, the Democratic Republic of Congo, Angola, South Africa, and Uganda that already have suffered Marburg [and/or Ravn]) virus outbreaks. By far, the most at-risk country where no detected modern outbreak has so far occurred is, by their findings, Ethiopia. The latter's stunningly high zoonotic niche population-at-risk score of 12,280,000 on Pigott and Golding's scale is nearly as high as that of the Democratic Republic of Congo, in the Sub-Saharan African MARV "hot zone," with a population-at-risk score of 13,140,000! Pigott and Golding's work was a somewhat more comprehensive and high technology version of earlier more narrowly mathematically based (using genetic algorithm for rule-set production or GARP) modeling and analysis by a team of University of Kansas scholars led by Dr. Townsend Peterson and Dr. Ryan Lash. Peterson and Lash's modeling correctly predicted the extremely deadly Uige, Angola Marburg outbreak in 2004. A delay in publication, however, somewhat diminished the impact of their novel and stunningly accurate prediction. Notably, no MARV risk factor in any of this research was found for Egypt despite the *Rousettus* population there. The University of Kansas team's modeling, however, also strongly indicated likely Marburg presence "particularly in Ethiopia" and centered in the Ethiopian Highlands. Pigott and Golding's work included graphs and maps summarizing their findings; the most relevant of these are reproduced above. A second 2017 article authored by a scientific team led by Dr. David Pigott, in the wake of the massive 2014–2016 West African Ebola outbreak, used similar modeling to chart outbreak potential across the African continent for four VHFs—Crimean Congo hemorrhagic fever (CCHF), Ebola, Lassa, and Marburg. Aimed at promoting continent-wide VHF outbreak preparedness, the included modeling again demonstrated high risk in areas considered "non-endemic" such as Cameroon and Ethiopia. In this second instance too, Pigott's modeling, regarding MARV, has received striking validation since being published. In 2022, there was an unprecedented fatal outbreak of Marburg virus in Ghana, right where Pigott's 2017 study predicted [23, 26].

The **human disturbance scenario** follows the same basic modeling lines and utilizes the same overlay of supporting evidence as the **anthropogenic attraction scenario**, with one key exception. The Ethiopian highlands *Rousettus* population, in this model, remains largely immobile, and rather successful, in adapting to the 240s CE drought. Expanding Aksumite civilization, pushing through the highlands from cities like Aksum, Coloe, and Adulis in search of ivory, gold, and emeralds for the Roman trade, however, disturbs the bats. Other accompanying practices such as deforestation (particularly notable in terms of its strong aforementioned relationship to mining and modern-day filovirus outbreaks in Africa), including felling/harvesting of fig timber, for a wide variety of purposes, also could have fragmented the bats' habitat and forced them away from more remote areas and into closer proximity to cultivated food sources [10, 14].

All three of these modeled scenarios have no obvious inferential/evidentiary issues. While we completely lack the evidence to be able to distinguish among them, their compelling cumulative overlay makes for another buttressing argument and congruent puzzle piece. This, on top of the original differential disease arguments, the varied direct and indirect evidence from the ancient sources, the climate pattern connection to VHF outbreaks, the contextual molecular-phylogenetic analysis of the members of the *filoviridae* and the supporting ecology of the Egyptian rousette bat [especially as the MARV reservoir host] that continues to build the case for the MARV evolutionary lineage as the cause of the Plague of Cyprian and enables us to contemplate the creation of a probable epidemiological model of the Plague of Cyprian. The latter is a task that we will undertake in Part III. Our resulting chain of analysis, modeling, and evidence has now finally brought us to the point of contemplating the likely mechanics of the necessary ancient zoonotic leap.

Likely Zoonotic Transmission Gateways—Kingdom of Aksum

While the precise interspecies disease transmission mechanics underlying many zoonotic pathogens (including the members of the *filoviridae*) are still not fully understood, it is generally accepted that around 75% of human diseases arose first in animals. Some better-known examples are influenza from waterfowl and bubonic plague from rats. SARS-CoV-2, the causal pathogen of the Covid-19 pandemic falls within a large clade of "SARS-related bat coronaviruses" identified in Asian bats. While some have suggested the virus may have leaked from laboratory studies at the Wuhan Institute of Virology, scientific consensus still rests with spillover from a wildlife "wet market," where any of the many live wild and domesticated animal species for sale could have served as a possible intermediate host connecting a putative bat source to humans. Even if a laboratory source was involved, the virus or a relative would have had to have been previously collected from an animal species in the wild. Of course, viruses are not the only pathogens making the leap from vertebrate animals to humans; bacteria, parasites, prions, and fungi can also cross over. Diseases generally produce little harm in their animal reservoir host populations but can be potentially devastating to humans. In modern times, the zoonotic transmission problem seems to be getting worse due to a combination of human disturbance causing biodiversity loss, even "biological annihilation" and the effects of globalization. In the twenty-first century, for example, 40% of all tick-borne vector zoonoses seem to have emerged in the last decade. A similar combination of worsening zoonotic disease factors can be seen at work in Roman times (though on a much smaller scale) along with a variety of different but similar amplifying epidemiological factors more unique to Roman antiquity that we will explore in Part III [1, 22, 38].

While a more detailed analysis of the transmission and progression of MARV will also be taken up in the following section, for now it is key to note that its onset generally begins between five to ten days after exposure. Although Marburg is not aerosolized or highly contagious and requires close contact for human-to-human transmission, the virus can be passed between humans by droplets, sweat, blood, saliva, feces, and urine. The virus can also be spread through sexual contact (a fact that emerged during the original modern outbreak in Germany in 1967) [30]. We know absolutely nothing concrete about the timing and epidemiological details of the Plague of Cyprian during its apparent initial outbreak in Aksum. We also know little about daily life in Aksumite cities or rural areas. We are also uncertain how deeply certain Roman practices (such as in medicine or socially) had penetrated Aksumite society or how much Romans knew about or were aware of Aksumite events. Thus, the use of any sort of epidemiological modeling along the lines of what we employ later in Part III for the Roman Empire becomes nearly impossible for Aksum. We can, however, note Aksumite, particularly from an ancient perspective, urbanization in an epidemiological context. This was on a far smaller scale than the Roman Empire's but still significant in context. There is, nonetheless, the problem of a complete lack of an Aksumite Cyprian or Dionysius/Eusebius.

We can potentially surmise that the plague in Aksumite cities frequented by Romans such as Adulis, Aksum, and Coloe was large enough to attract the attention of ancient Greco-Roman commentators (now vanished sources upon which the later Greco-Roman, and, even later, Byzantine commentators cited by Harper relied). Though, conversely, they were also potentially entirely ignorant of it. We will discuss a quite plausible scenario for the ancient "Ethiopian" origin evidence that assumes that the Romans were, in fact, totally unaware of the, highly probable in our view, Aksumite origins of the Plague of Cyprian. This analysis again, inherently accepts the verdict of those sources and, for reasons stated earlier, does not accept Huebner's characterization/dismissal of that evidence through the explanation of a xenophobic Greco-Roman *topos*. The potential highly uncertain further inference that the importance of this exotic event might have loomed larger after the devastating spread of the pestilence to the Roman Empire, however, also creates a yet additional element of uncertainty both in terms of interpretation and in terms of the potential scale of the Aksumite outbreak. In the current section and in Part I, we believe that we have amassed considerable evidence to support and buttress Kyle Harper's original conclusion that the Plague of Cyprian probably erupted beyond Rome's southeastern frontier. This would, as we have demonstrated, certainly highly likely mean Aksum. Yet, getting beyond that point analytically with anything approaching coherence and high probability in our modeling was very challenging due to the considerable ancient evidentiary challenges, gaps, and silences regarding Aksum. In response, we developed a specific and highly complex multifaceted overlay of the range of available ancient and modern multidisciplinary evidence from our Parts I–II. This analysis, admittedly, builds strongly on the basis of the conclusions that we have developed so far. The holistic synthesis of all this evidence is summarized in Table 6.1.

An urban pestilence in Aksum, whether noted by Roman commentators or not, that subsequently spread to Roman territory would have likely done so during the peak annual maritime trading "inflow" (from a Roman perspective) season given the documented intense level of inbound trade contacts. It is also nearly certain, epidemiologically, to have been disseminated during that trading season window rather than during the remaining months of the year. Inbound Roman-Aksumite contacts would have presumably sharply diminished during that "off-season" period (especially during the *mare clausum*), and the maritime flow at other times of year would have been largely outbound *to* Aksum. The following findings do suffer somewhat from an irresolvable problem of partial conflation between the timing of the Aksumite outbreak itself and the timing of its dissemination to the Roman Empire. There are also certain unavoidable imprecisions, circularities and minor flaws. Yet, the ultimate temporal indication of the table is also, ultimately, surprisingly (in light of its overall complexity) unitary and overwhelming.

The epidemiological timeline for an Aksumite outbreak in early 249 CE, seemed to us, given the evidence presented in Part I, simply too short. Additionally, since trade and movement would have been at a very

Table 6.1 Calculating TOW/DS

	Months of the year, 248 CE											
	Jan	*Feb*	*Mar*	*April*	*May*	*June*	*July*	*Aug*	*Sept*	*Oct*	*Nov*	*Dec*
Data in support of MARV outbreak & subsequent dissemination												
Tractate Taanit (potential drought-breaking dry to wet temporal parameters)			X	X					X	X		
Deduced from Dionysius' chronology											X	X
Belg/Kremt (relevant climate/rainfall patterns)		X	X	X	X	X	X	X	X			
Nile flooding (traditional)							X	X	X			
Seaborne trade (window—near side)						X	X	X	X			
Seaborne trade (return window—near side)									X			
Seaborne trade (window—far side/India)							X					
Seaborne trade (return window—far side/India)									X	X		
Latter half of 248 CE weighted more heavily due to the proposed timing of the Alexandrian outbreak and the proximate linkage to plague in "Ethiopia" in the ancient sources							X	X	X	X	X	X
Dry to wet season transition period in sub-Saharan Africa (linked to MARV pulses in bats, nonhuman primates; measured with satellite data)			X						X			
Plague of Cyprian seasonality	X	X	X						X	X	X	X
No mare clausum (mediterranean sea lanes open)				X	X	X	X	X	X	X		
Favorable red sea currents/winds	X	X	X	X					X	X	X	X
Augmented score factors												

(continued)

Table 6.1 (continued)

	Months of the year, 248 CE											
	Jan	*Feb*	*Mar*	*April*	*May*	*June*	*July*	*Aug*	*Sept*	*Oct*	*Nov*	*Dec*
1st Juvenile Egyptian Rousette Bat Marburg Pulse—0.5/month values	X	XX	XX									
2nd Juvenile Egyptian Rousette Bat Marburg Pulse—0.5/month values								XX	XX	X		
Total outbreak window/dissemination score (TOW/DS)—base/augmented	2/3	3/5	5/7	5/5	2/2	3/3	6/6	5/7	11/13	6/7	4/4	4/4

Table by Dr. Mark Orsag

low ebb at that time, dissemination would also have been very unlikely. We therefore concentrated on 248 CE in our modeling in terms of overlaying the modern and ancient evidence that we do have (…an even earlier Aksumite outbreak date seems, in light of the considerations above, at least somewhat unlikely—as no such outbreak appeared to have affected the strongly commercially linked Roman Empire prior to 249 CE). After all, as we argued in Part I, the plague seems to have arrived and built to a peak in Alexandria by the early spring of 249 CE. The epidemiological spread and urban "slow burn and build to peak" timelines for Alexandria, Carthage, and the Roman Empire as a whole seemed to have ranged, roughly, from around 5–7 months for Alexandria to around 6–7 months for Carthage, to around 2.5 to 4 years, in cascading fashion, for the Empire as a whole. Given all the evidence from Parts I–II so far, supplemented by additional modern scientific evidence, we believe that it is possible to calculate a probable broad temporal window for the initial Aksumite outbreak/subsequent dissemination to Roman territory within the year 248 CE. There is some admittedly slightly troubling circularity necessarily inherent in this rather inferential result, but the holistic interdisciplinary analysis embodied in the table/model as a whole achieves enough of a broader multifaceted fit and multidirectional consilience that we did not find these problems of circularity to be discrediting overall.

Our resulting, rather complex and admittedly somewhat speculative, calculations also seem somewhat more convincing for the timing of epidemiological pathogen dissemination to the Roman Empire, as opposed to a temporal pinpointing of the initial Aksumite disease outbreak. The latter is far more difficult to determine precisely due to a complete lack of applicable ancient Aksumite source material. We have no real way, for example, to model the potential epidemiological socio-behavioral R0-influencing environment for an Aksumite outbreak whatsoever, beyond noting the presence of three Roman-influenced small to medium-sized urban epidemiological environments. Yet, the overall analytical framework that we have applied also suggests that the two events may well have been at least relatively proximate in time. In fact, the Roman outbreak, given the relatively slow burn to build times noted and modern experiences and modeling of MARV outbreaks, may well have been quietly underway by the time ancient Romans noted the Aksumite one (though it is also possible, as we noted earlier, that they never were cognizant of it). Our overall findings are summarized in Table 6.1—Calculating TOW/DS: 248 Outbreak Window (Aksum)/Dissemination (to Roman Egypt). The table displays temporal indicators, in which we place the likely occurrence of [positive or likely = X, negative or unlikely = (blank)] 14 different types of evidence into each of the months of 248 CE. In the table's first (base) calculation of a Total Outbreak Window/Dissemination Score (TOW/DS), September 248 CE, scores by far the highest (with a total base score of 11 out of 14 possible analytically contributing evidentiary points valued as positive or likely). If one puts some further potential for circular/reductionist reasoning aside and includes the Egyptian rousette bat bi-annual juvenile pulse windows in the calculation (augmented score), September scores 13 out of 16 possible points. October (when Red Sea current patterns would have become annually directionally fully favorable for trading voyages) was one of two months (along with July) in a tie for second—scoring 6 (non-augmented) base points—thus creating a very strong potential "outbreak/dissemination window" for the early to mid-Fall of 248 CE.

Despite this modeling/analysis, the overall picture is still incomplete. The whole timeline regarding the outbreak in "Ethiopia" is far from certain due to a near total lack of *specific* evidence relating to the outbreak and progress of the Plague of Cyprian at its apparent Aksumite point of origin. Yet, the Fall 248 CE probability calculation is, however, more fully convincing as a temporal marker for potential dissemination to Roman territory. The viral pulse windows are, however, when 83% of modern zoonotic transmissions/human outbreaks of MARV begin; the nearly complete congruence with the second augmented score temporal window here is also interesting if far from definitive [3]. A rapid transfer of the pathogen to Roman territory given the overall high volume of trade contacts can't be excluded but neither can an Aksumite outbreak somewhat earlier in the year. For example, during the 1st January–March JERB MARV window. That would have given the outbreak time to build in Aksum and *later* disseminate to Roman territory via Red Sea-based maritime commerce during the early fall commercial peak. To us, the latter scenario seemed somewhat more likely. The months-long extended "slow burn-and- build" modeled MARV outbreak patterns that we will discuss in Part III, however, increase rather than diminish potential the problem of conflation. Even some of the table's flawed results still contribute to consilience in terms of the indicated overall result. Row 3, for example, notes likely Alexandrian outbreak presence in November and December. Yet there would have been, as we discuss in Part III, two to three intervening/cascading/overlapping outbreaks between Aksum and Alexandria. Those, of course, would have been underway in the months before November, which again points strongly to the September–October window, Overall, the patterns embodied in Table 6.1 also achieve a complex congruence/consilience with Rathbone and Harper's analysis that the year 248 CE marked the end of the worst of the 240s CE regional drought. There is also alignment with the drought-breaking windows that we noted in the *Tractate Taanit* in Part I, with the peak of inbound maritime trade from the Red Sea to the Mediterranean (before the initiation of the *mare clausum*), with the end of the Nile flooding season and with the 2nd juvenile pulse window of the annual birth and peak Marburg infectiousness windows of the ERB.

Some partial potential circularity to the whole complex analysis, despite the strong multidirectional consilience achieved, also made us slightly uncomfortable. Therefore, the split into separate TOW/D. score sets in Table 6.1; one factoring in the pulse windows as evidence (augmented score) and one deliberately excluding them (base score). Inclusion of the augmented TOW/DS points creates a three-way tie between October, August, and March (7 points). September 248 CE, however, still scored a runaway 13 out of 16 points. The values for September 248 CE in terms of both base and augmented scores, and the overall clear overlap of probability with the 2nd juvenile viral pulse window are, despite some of the real potential analytical/modeling conflation and circularity problems and complications, both notable results. Yet, the consistently very high outlier base and augmented scores for September 248 CE, in particular, also provide at least slight further overall congruence linking the Egyptian rousette bat and the MARV evolutionary line to the Plague of Cyprian. Such values strongly, with far fewer analytical caveats, support a modeled September–October 248 CE dissemination window for the spread of the plague from Aksum to Roman Egypt. It would then peak in Alexandria approximately eight months later. The overall evidentiary overlay is strongly indicative and consistent, but due to the overall extreme complexity of the predictive modeling here and the aforementioned potential for conflation, circularity, and reductionism admittedly well short of fully revelatory. The pieces seem to fit, but there are still a number missing or rather uncertain in this particularly thinly evidenced, and therefore particularly challenging, section of the Plague of Cyprian puzzle.

Contrastingly, there are several clear and viable potential paths of zoonotic transmission applicable to the situation that prevailed in 248

CE, as we have modeled it, in Aksum. The demonstrated modern pattern of MARV, however, is notable (and we felt telling) in this regard. "Multiple introductions of the virus from the same zoonotic pool" were demonstrated in the mining-related Durba outbreak in 1998–2000 in the Democratic Republic of the Congo. The strongly cavernicolous nature of the Egyptian rousette bat also provides additional supporting evidence in regard to MARV lineage implication and a likely zoonotic transmission pathway through mining activities. This type of scenario would also provide a potentially much larger group of initial infections than postulation of ancient isolated Karst cave explorers. This is an important factor given the rather epidemiologically fragile and highly stochastic beginnings of a potential MARV outbreak. A modern parallel would be the Durba mine (128 deaths) vs. the Python Cave (1 death) outbreaks of MARV. This supposition is given additional support by the ancient–modern overlay of possible intensive deforestation in the vicinity of ancient mining operations and the aforementioned "strong association" of the ecological disturbance caused by deforestation with modern filovirus outbreaks. This overall scenario would also have given the brewing pandemic a broader more sustainable momentum at its crucial, and epidemiologically fragile, beginnings. Members of the African *filoviridae*, being non-airborne pathogens, generally have very modest base biological R0 "infection spread" component values [this terminology is, however, complicated and will be discussed more thoroughly in Part III]—often, if far from always, above the minimum 1.0 level (necessary for outbreak sustainability) but nearly always below 3.0 (this in advance of the application of any modern medical epidemic-mitigation measures). So, such an initial "snowball effect" would have been important in preventing a rapid end to the spread of the deadly pathogen. The annually almost certainly high volume of Red Sea trade in September, with the seaborne trade about to at least partially shut down for the winter *mare clausum* (November–March) in the Mediterranean, between Aksum (and beyond) with the Roman Empire, might well also have led to increased contacts between Egyptian and Greco-Roman merchants and Aksumite miners, ivory merchants, and other traders during the Fall of 248 CE. They would likely have interacted in the bustling, more densely populated Aksumite urban trading environments on established trade routes (Adulis, Coloe and Aksum) that were infectiousness-enhancing in an epidemiological sense. Overland travel times involved were also relatively short and, as noted earlier, all well within the average infection window for MARV (minimizing the potentially very strong epidemiological frictions created by ancient patterns and means of trade and movement) [5, p. 19; 20, 25, 31].

There are two other alternative zoonotic paths that cannot, however, be dismissed. They thus require some detailed examination in light of both modern concerns and ancient practices. The first concerns potentially contaminated fruit—such as figs. "Any fruit the [infected] bat bites can be smeared with Marburg [in the ERB's saliva]; a person, monkey, or other animal eating the fruit can get infected." So concluded twenty-first-century CDC scientists, who were so concerned about the deadly Marburg virus, in several contexts, including food-chain contamination that they sought funding from the Pentagon's Defense Threat Reduction Agency. They indeed later demonstrated that MARV could persist on discarded fruit for hours. The threat of such immediate local contamination could have been and indeed is today very real, particularly during the bi-annual birth pulses, given *Rousettus'* regular fifteen-mile nightly feeding-dispersal range and potential use of multiple roosts. One absolutely can't rule this scenario out as a potential Plague of Cyprian zoonotic origin point. Consumption, however, would have to have been immediate as the ancient drying processes used for fruits like figs or dates (both within the Roman Empire and Aksum) would have killed the virus. The feeding of contaminated fruit (possibly unripened if the feeding habits of *R. aegyptiacus* are fully considered), to animals also isn't a terribly plausible zoonotic secondary gateway in this regard. None of the animals commonly cultivated in ancient Aksum—goats, sheep, or cattle have shown any sign ever of being infected with or carrying/shedding filoviruses. Cooking also kills the virus. According to the findings of a team of Russian virologists at the leading "Vektor" laboratory, sheep and goats also indeed seem to be "completely insensitive" to members of the *filoviridae*. Pigs are in another category altogether (and there is pig husbandry in modern Ethiopia), but there is simply zero evidence reliably placing swine as part of third-century Aksumite animal agriculture. The Roman Empire itself was a quite different matter in the porcine department that we will cover in Part III. There is also no evidence that the Roman Empire, which grew its own figs in abundance, imported figs from Aksum or that the virus could have survived the travel and treating processes involved. Contacts between Aksumite agriculturalists (unlike miners or ivory merchants) and Roman traders are completely undocumented in the ancient sources. And yet, Roman traders would have had to eat while in Aksum and may well have bartered with local growers [4, 13, 31, 32].

The existence of a trade in ERB body parts for Roman folk medicinal purposes or bats kept as pets is also at least a reasonable possibility—especially given that Rome was a major trading partner of Aksum. Roman folk medicine practices might well have created a potential market for ERB organs or tissue. *Rousettus* abundance, attraction to anthropogenic environments and status as an agricultural pest would have made it a target, at least by inference, for such harvesting of bat tissue or organs. Again though, while eminently possible with direct animal handling and butchering in a local Aksumite sense, such a scenario seems speculative and plausibly made less likely by the application of unknown but clearly necessary ancient harvesting and preservation practices. As with the example of figs, the Egyptian ERB population would have given Roman merchants and exotic animal dealers a viable internal resource in this regard. The epidemiological basis of such a folk medicine/exotic animal dealer driven outbreak, like one involving consumption of fresh fruit, would also probably have been narrower and less sustainable than one initially involving multiple infected miners; initial epidemiological fragility and stochasticity apply here too. Preparation/preservation, distribution, and usage practices for bat parts in Roman folk medicine are also unknown. Quite probably, the virus might also have been rendered inert by them. Yet, those carrying them out would have risked exposure in a modern epidemiological parallel to "bushmeat related activities" as opposed to consumption [2, 15, 16, 33, 34].

Under the **human disturbance scenario**, Aksumite mining operations, given the overlapping patterns of scientifically studied modern MARV outbreaks and the likely epidemiological mechanics, nature, and annual timing of the thriving trade/commercial relations between Aksum and the Roman Empire, constitute the most probable pathway for the Aksumite zoonotic leap that later snowballed into the Plague of Cyprian. Serological evidence (albeit less dispositive than the gold standard of virus isolation from multiple individuals) shows that miners, in an unsurprising and almost certainly timeless broader correlation, are consistently *around 5.4 × more likely* to be seropositive (to have detectable antibodies to a filovirus) than the surrounding general population. This from a 2020 study conducted in Ugandan regions that had witnessed multiple Marburg and Ebola outbreaks. The ancient evidence of known Ethiopian highlands-mined exports of gold, emeralds, and iron from Aksum to the Roman Empire completes this supporting modern–ancient overlay neatly in an analytical sense. The ancient capital city of Aksum was located in the northern Ethiopian highlands. Another commonly traded commodity that would be somewhat excluded biogeographically but potentially brought into play under the aforementioned **anthropogenic attraction scenario** would be salt. Similarly, this anthropogenic model allows for more direct attraction of the bats to these Aksumite cities and the agricultural lands around them (this possibility finds something of a partial modern epidemiological analogue in the case of the large Uige, Angola MARV outbreak detailed below). As will be discussed in Part III, the further overlay of modern epidemiological evidence in this regard also very strongly reinforces this conclusion. It is obviously impossible to precisely pinpoint potential locations, and distinguish definitively (though a zoonotic outbreak among miners appears by far the most probable pathway) between these three third-century zoonotic/epidemiological scenarios. Even in well-documented twenty-first-century MARV outbreaks, initial zoonotic pathways and sometimes precise outbreak timing are often unclear [8, 19, 24].

There is no more dramatic example than the deadly Marburg outbreak that terrorized the Uige region of Angola beginning in early October 2004. The outbreak was finally resolved after about one year. It killed hundreds of people, and the disease was nearly 90% fatal. The outbreak took place in a disordered violent region that was still in the aftermath of a recent civil war. While the detection date of the initial cases in early October 2004 would likely indicate that, as usual, *R. aegyptiacus*' pattern of bi-annual birth pulse-related infection was probably responsible, no clear initial zoonotic pathway emerged. The area of the outbreak was far more urban than anthropogenically rural. No free-standing fruit groves were nearby and no local *Rousettus* colony was discovered. The main *foci* of the outbreak were cases of human-to-human transmission in home care settings and local and regional health care facilities where the disease was transmitted through insufficiently sterilized medical equipment (or other disinfection protocol violations) used unofficially in private homes or officially in hospitals displaying "unsafe practices that amplify transmission in a health care setting." One pediatric hospital ward in Uige was a particularly notable and horrifying epicenter, and the disease "slowly burned and built" to a nearly unremittingly fatal peak by the spring of 2005 [18].

Once the MARV evolutionary lineage pathogen got into Aksum's relevant urban population(s) in a more extensive way, and given the modern average incubation period of 5–10 days (or even in certain rare cases as long as 21 days), there is no doubt that the virus could easily have breached the imperial frontiers through ongoing Aksumite-Roman Egypt trade contacts [17]. As noted, with a looming return date/seasonal cutoff for shorter-range and longer-range seaborne trade routes, the volume would probably have been particularly heavy in the month of September 248 CE. The virus would have traveled along the relevant seaborne trade routes. Sea (and by water more generally) is a much more probable epidemiological vector than land in this third-century case, given the notably greater volume, cost efficiency, and particularly speed and relative physical ease of movement. Aksum's road systems were also not equivalent to those of the Roman Empire. The overland movement would thus have been even slower and more epidemiologically problematic outside the empire than within it—given the interrelated factors of time, distance, rugged terrain, incubation periods, and severe victim debilitation/high mortality rates. Taking all these factors into account, the infection would have very likely failed to spread efficiently over longer overland distances, even when supplemented with riverine travel up the Nile.

Red Sea currents are weak and variable. Other than near the mouth where weak northwesterly winds (unfavorable in our context) prevail, wind patterns are also highly seasonably variable. The key direction in terms of the modeled viral epidemiology is south to north (Adulis to the main Roman Red Sea entry ports of Berenice and Myos Hormos). Seaborne trade moving in that direction would have encountered unfavorable wind and current conditions between May and August. September to April would have seen generally more favorable Red Sea wind and current patterns but shipping volume would also likely have been affected highly negatively by the *mare clausum* [21]. The month of April in the spring and the months of September (yet again) and October seem to be the most likely windows. The virus would have, after having breached the empire's boundaries by sea, then moved secondarily by land/river toward Luxor, Coptos, and Alexandria, exactly as the Plague of Cyprian did according to Harper's convincing analysis of the ancient sources, southeast to northwest into the Roman Empire. The major cities of Roman Egypt—Thebes/Luxor, and Alexandria (we will discuss these overland and riverine travel times in Part III), as well as the potential key entry ports of Berenice and Myos Hormos all could have been affected early and heavily. From the major hub of Alexandria in particular, the virus would have spread, primarily (if not necessarily solely) through seaborne and riverine trade and movement, steadily throughout the Empire.

While we have no surviving direct evidence regarding the latter two listed locations, there is very strong evidence for **both** of the formerly listed locations. Although some pieces are still missing, those that are there seem to fit perfectly once again. Red Sea sailing times under such generally favorable conditions in the likely fall window (September–October) wouldn't have been, by our calculations, prohibitively long either—around 1.5 to 3 days from Adulis to Berenice, and 6 to 8 days from Adulis to Myos Hormos. Such values are all well within the average exposure/infection window of 5–10 days for Marburg. Infections aboard ships on such relatively short voyages don't pose the same insuperable issues that apply to longer, in a temporal sense, and additionally physically arduous, overland travel. Whether already clearly sick or yet to show symptoms, individuals coming or being brought ashore at Berenice or Myos Hormos after sailing from Adulis could very plausibly not only have been the source of the plague but also could constitute one highly plausible reason (though not the only one as we will explore in Part III) that ancient sources might have linked the pestilence's origins to "Ethiopia." While in these port towns, those infected might well have interacted with the rather ubiquitous urban Roman institutions of the medical profession, the baths, and (if they died) the third century's elaborate funeral practices…more on all those in Part III. The port city of Adulis, which Pliny had once described in the first century as the principal commercial center of "Ethiopia," would have easily been the shortest and most likely high traffic point of origin/entry in non-Roman Africa for the Plague of Cyprian. The maritime trade routes between it and Berenice and Myos Hormos constitute the proverbial shortest, easiest, straightest, and most epidemiologically sustainable disease dissemination nexus between Aksum and Roman Egypt. These calculations and travel times are deduced from values contained in the *Periplus*, annual (admittedly somewhat variable) Red Sea current and wind patterns, and the average sailing speeds of Roman merchant vessels. Travel times apply to the traditional maritime "return window" in the fall. As noted, more generally favorable wind and current patterns prevailed in the Red Sea and beyond (to India) and made such return trips easier and more practical. This was the reason, along with the looming *mare clausum* (partial [ancient sailors could and occasionally did set sail in winter if they had a very compelling reason] or full closing of the Mediterranean to sailing during the winter months), for the annual September–October pattern of inbound (returning) seaborne trade [7, 9, 12, 21, 27–29].

The holistic overall evidentiary "systems-synthesis" congruence and consilience holistically achieved through the first two installments of our analysis is, we believe, quite compelling. There are still, however, many remaining questions for Parts III and IV. How did the outbreak likely function epidemiologically and spread locally, regionally, and more universally in the third-century Roman Empire? What were the mechanics of infections? How did the Plague of Cyprian likely function across the range of epidemiological settings that existed in the third-century Roman Empire? In what ways would an outbreak of MARV or one of its ancestors in the Roman third century have looked similar or different in size, duration, and scale to twentieth and twenty-first-century Sub-Saharan African MARV outbreaks? Or to put it a different way, could a very deadly but non-aerosolized and not terribly biologically contagious pathogen such as MARV have caused an ancient pandemic event like the Plague of Cyprian? In the twentieth and twenty-first centuries, the biggest MARV outbreaks have killed hundreds of people, not thousands or tens of thousands. They have lasted one to two years, not fourteen or fifteen. What would have been different about a major third-century Roman MARV lineage outbreak? What can we deduce, model, and recreate about the nature, functioning, and likely impact of the Plague of Cyprian pandemic in terms of its pathology, epidemiology, and history? Can we potentially determine some of its broader effects and even its broader potential contribution to the Roman Empire's Third-Century Crisis? Or even, its contribution to the empire's eventual complex "Fall"? Finally, what broader lessons, if any, are there for a twenty-first-century world still reeling from the effects of the Covid-19 pandemic?

References

1 Aguirre, Alonso, 2017 "Changing Patterns of Emerging Zoonotic Diseases in Wildlife, Domestic Animals and Humans Linked to Biodiversity Loss and Globalization," *ICRR Journal Oxford Academic*, 12/15/2017 https://doi.org/10.1093/ilar/ilx035.

2 Ajelli, M. and Merler, S. 2012 "Transmission Potential and Design of Adequate Control Measures for Marburg Hemorrhagic Fever," *PLoS One* https://doi.org/https://doi.org/10.1371/journal.pone.0050948.

3 Amman, B.R., Carroll, S.A., et al. 2012 "Seasonal Pulses of Marburg Virus Circulation in Juvenile *Rousettus aegyptiacus* Bats Coincide with Periods of Increased Risk of Human Infection," https://www.ncbi.nlm.nih.gov/pubmed/23055920/. Accessed on 0422/2020.
4 Amman, Brian, et al. 2021 "Marburg Virus Persistence on Fruit as a Plausible Route of Bat to Primate Filovirus Transmission," *Viruses* 13: p. 2394. https://doi.org/https://doi.org/10.3390/v13122394
5 Bardi, Ugo, 2018. *The Seneca Effect* (Cham, Switzerland: Springer).
6 BBC, 2022 "Ghana Confirms First Cases of Deadly Marburg Virus," http://www.bbc.com. Accessed on 7/19/2022.
7 Beresford, James, 2013 *The Ancient Sailing Season* (Leiden, Brill).
8 British Museum, 2013 "The Wealth of Africa – The Kingdom of Aksum" *britishmuseum.org*. https://www.yumpu.com. Accessed on 5/17/2021.
9 Casson, Lionel, 1951 "Speed Under Sail of Ancient Ships," *Transactions and Proceedings of the American Philological Association* 82: pp. 136–148. https://doi.org/10.2307/283426
10 Centeno-Cuadros, A. and Hulva, P., 2017 "Habitat Use But Not Gene Flow Is Influenced in Two Ecotypes of Egyptian Fruit Bat (*Rousettus aegyptiacus*)", *Molecular Ecology,* https://doi.org/10.1111/mec.14365.
11 Cyprian, "An Address to Demetrianus," *The Complete Works of Saint Cyprian of Carthage*. Translation Editing: Nicene Fathers [1885] Philip Campbell, Ed. Introduction by Ryan Grant 2013 (Evolution Publishing): pp. 93–108.
12 East, Gordon W. 1938 *The Geography Behind History* (W.W. Norton), pp. 174–175.
13 Gonchar, N.I. and Pschenichnov, V.A., 1991 "The Reactivity of Different Experimental Animals to the Marburg Virus," *Vektor* [in Russian] PMID: 1803784.
14 Harkness, Hannah and Tefera, M. 2004 "Agriculture in Aksum" and "Recent Evidence of Animal Exploitation in the Axumite Epoch, 1st-5th Centuries," *Tropical Animal Health and Production*. https://doi.org/10.1023/B:TROP.0000012100.96577.5f. Accessed on 04/27/2020.
15 Kudoyarova-Zubavichenie, N.M. et al., 1999 "Preparation and Use of Hyperimmune Serum for Prophylaxis and Therapy of Ebolavirus Infections," *State Research Center of Biotechnology*. https://www.watermark.silverchair.com. Accessed on 6/11/2021.
16 Kurpiers, Laura, et al., 2015 "Bushmeat and Emerging Infectious Diseases: Lessons from Africa," in F.M. Angelici, ed., *Problematic Wildlife: A Cross-Disciplinary Approach* (Cham: Springer): pp. 507–51. https://doi.org/10.1007/978-3-319-22246-2_24.
17 Mayo, "Ebola Virus and Marburg Virus, 2016," Mayo *Clinic*. https://www.mayoclinic.org. Accessed on 8/9/2020.

18 Ndayimirije, Nestor, 2005 "Marburg Hemorrhagic Fever in Angola—Fighting Fear and a Lethal Pathogen," *New England Journal of Medicine*. https://doi.org/10.1056/NEJMp058115.

19 Nyakarahuka, Luke, Schafer, Ilana, et al., 2020 "A Retrospective Cohort Investigation of Seroprevalence of Marburgvirus and Ebolavirus in Two Different Ecological Zones in Uganda," https://pubmed.ncbi.nlm.nih.gov/32611400/. Accessed on 7/5/2022.

20 Olivero, Jesus, John, E., et al., 2017 "Recent Loss of Closed Forests Associated with Ebola Virus Disease Outbreaks," *Scientific Reports/Nature Research*. https://doi.org/10.1038/s41598-017-14727-9.

21 Patzer, W.C., 1974 "Wind-Induced Reversal in Red Sea Circulation," *Deep Sea Research*, 21(2): pp. 109–121.

22 Pekar, Jonathan, et al., 2022, "The Molecular Epidemiology of Multiple Zoonotic Origins of SARS-CoV-2," *Science* 0(0): eabp8337 https://doi.org/10.1126/science.abp8337.

23 Peterson, Townsend and Lash, Ryan, 2006, "Geographic Potential for Outbreaks of Marburg Hemorrhagic Fever," *American Journal of Tropical Medicine and Hygiene* 75(1): pp. 9–15. American Society of Tropical Medicine and Hygiene. https://kuscholarworks.ku.edu/buitstream/handle/1808. Accessed on 4/24/2020.

24 Phillipson, D.W., 1998 *Ancient Ethiopia: Aksum, Its Antecedents and Successors* (British Museum Press).

25 Pigott, David, Golding, Nick, et al., 2015 "Mapping the Zoonotic Niche of Marburg Virus Disease in Africa," *Transactions of the Royal Society of Tropical Medicine and Hygiene*, 3/27/2015. https://doi.org/10.1093/trstmh/trv024.

26 Pigott, David, et al., 2017 "Local, National, and Regional Viral Hemorrhagic Fever Pandemic Potential in Africa: A Multistage Analysis," *Lancet (London, England)*, 390(10113): pp. 2662–2672. https://doi.org/10.1016/S0140-6736(17)32092-5.

27 "Roman Ships and Navigation in Ancient Rome," 2018. https://www.vita-romae.com. Accessed on 6/4/2022.

28 Schoff, Wilfred, 1917 "Navigation to the Far East Under the Roman Empire" *Journal of the American Oriental Society* 37: pp. 240–249. https://www.jstor.org/stable/592924?seq=1#metadata_info_tab_contents. Accessed on 10/18/2020.

29 Schoff, Wilfred, trans. 1912 *The Periplus of the Erythraean Sea; Travel and Trade in the Indian Ocean*, "The Voyage Around the Erythraean Sea," pp. 22–23 (1–8). https://archive.org/stream/cu31924030139236/ Accessed on 4/20/2020. [Inclusive].

30 Slenczka, Werner and Klenk, Hans Dieter, 2007 "Forty Years of Marburg Virus," *The Journal of Infectious Diseases* 196(Supplement 2): pp. S131–S135.
31 Stein, R.A., 2014 "What Is Ebola," *The International Journal of Clinical Practice* https://onlinelibrary.wiley.com/doi/full/10.1111/ijcp.12593 Accessed on 1/23/21.
32 Sun, Lena, 2018 "On a Bat's Wing and a Prayer," *The Washington Post.* https://www.washingtonpost.com. Accessed on 8/13/2020.
33 Tommasi-Moreschini, Chiarra O., 2011 "Bats in Greco-Roman Antiquity," *BATS Magazine: Media and Education* 29(2). https://batcon.org/article/bats-in-greco-roman-antiquity/ Accessed on 7/20/2020.
34 WHO, 2014 "Ebola and Marburg Virus Disease Epidemics: Preparedness, Control, Alert, Control and Evaluation". https://www.apps.who.int. Accessed on 09/20/21.
35 WHO, 2021 "Fact Sheet-Marburg Virus Disease." https://www.who.int. Accessed on 2/23/2022.
36 WHO, 2021 "Guinea Declares End of Marburg Virus Disease Outbreak." https://www.afro.who. Accessed on 12/21/2021.
37 WHO, 2019 "Marburg Virus Disease." https://www.who.int. Accessed on 12/21/2021.
38 Worobey, Michael, et al., 2022 "The Huanan Seafood Wholesale Market in Wuhan Was the Early Epicenter of the COVID-19 Pandemic," *Science*, 0(0): abp8715. https://doi.org/10.1126/science.abp8715

PART III

"The Why": Projected MARV Lineage Epidemiology and Pathology in the Third Century Roman Empire

"[Meaning]...that the virus was taken from Philadelphia to San Francisco, then to New Orleans, Rio de Janeiro, Rome, Karachi, Bangkok, and then Beijing."—James Cole—*12 Monkeys (Directed by Terry Gilliam)*, 1995.

CHAPTER 7

Guardrail Modeling: Geographical Dissemination Pathways and the Urban Epidemiological Setting

Part III is, in many ways, the decisive section of convergence for our holistic solution; our purpose is, therefore, multifold. Modeling, as it does in a generalized "guardrail" sense, the "why" of the Plague of Cyprian and building an understanding of the plague's likely functional epidemiology throughout the Roman Empire is our focus. Our epidemiological modeling of the Plague of Cyprian is, of course, informed by our retrospective diagnosis conclusions regarding the MARV evolutionary lineage pathogen and also by the likely zoonotic origin, timeline, and dissemination pattern outlined in Part II. Additionally, the conclusion of Part II left us with the initial task of tracing the likely path of the virus once it breached the imperial borders and, in doing so, checking the congruence of our detailed likely disease dissemination trajectory with Kyle Harper's original Plague of Cyprian epidemiological timeline and directionality, as detailed in Part I.

It is at this point where we explicitly accept the results of our retrospective pathogen diagnosis in Part II and use the modern example of Marburg lineage viral pathology and epidemiology as part of our overall modeling equation going forward. Again, the key test here will be the system's synthesis analytical fit with what we independently know from the ancient evidence. While we will discuss other elements of MARV's epidemiology and pathology a bit later in this section, one immediate focus is on the infection window of roughly 5–10 days after exposure.

M. Orsag et al., *Interdisciplinary Insights from the Plague of Cyprian*,
https://doi.org/10.1007/978-3-031-26094-0_7

Another scientific expression that we will utilize in this regard is Tg (generation time). This is defined technically as the temporal lag between infection in a primary case to infection in a secondary case epidemiologically deriving from the primary case. While these two temporal measures aren't completely equivalent in a medical sense, we will use them somewhat interchangeably for our "guardrail modeling" purposes here. MARV's "relatively long" Tg is roughly 9 to 9.5 days. For modern epidemiologists, such "knowledge is required to manage a potential Marburg hemorrhagic fever epidemic." At a basic epidemiological level, MARV isn't aerosolized or terribly infectious with a "relatively small reproduction number" that was calculated as being approximately 1.59 [one infected person would, on average, infect 1.59 others] during the largest modern outbreak. Yet it is one the "most virulent pathogens infecting humans with CFRs in major outbreaks even higher than 80%" [2].

Modeled Spatial Epidemiology of the Maritime Environment in the Roman Third Century

These numbers loom particularly large for us due to the fact that Roman commercial sailing vessels *would not* have been infectiousness-enhancing environments in an epidemiological sense, and MARV's basic biologic contagiousness is, after all, fairly low. It is easy to imagine the dark hold of an ancient ship and speculate differently, but it wouldn't have been so. There were no passenger ships in the third century Roman Empire as such. Travelers booked passage on merchant ships. Many of these were quite large and spacious. An average Roman merchant vessel had an impressive carrying capacity of 100–150 tons. Larger "grain supertankers" like the exemplar of the *Isis* had cargo-carrying capacities of up 1,200 tons and were up to 55 meters (well over half an American football field) long! No ships of this size were built again after the Roman Empire's collapse until the sixteenth century. Massive twenty-first-century *Panamax*-type container ships, for comparison, are around 290 meters long and have 44 times the carrying capacity of a Roman *Isis*-type merchant vessel. In all but the most extreme weather circumstances, crews and passengers on Roman civilian and military ships spent most of their time (including sleeping at night) on deck and during the day (at times exposed to "nature's disinfectant"—sunlight). Notably, in an epidemiological context, passengers and crew brought, or were issued, their own personal “food supplies, covers and mattresses.” Tents were sometimes pitched on deck as well. While we know less about crew compositions and sizes, cost concerns would likely have kept those as small as practically possible. The wealthiest class of Romans seeking passage might negotiate a price for a small above deck cabin usually located near the ship’s stern. Some military designs did feature rowers who operated in tightly confined spaces below deck, but they weren’t confined there, contrary to popular perception. They also generally weren’t slaves, but rather either soldiers themselves or a specialized class of freedmen from the provinces. We did not model, however, the epidemiological environment prevailing in the mid-third-century Roman navy. There were occasionally confined slave oarsmen on merchant ships, but these ships usually relied far more, due to cost concerns, on wind power [7, 41].

Some of the assumptions in our modeling require some elucidation here. Particularly after awareness of the pestilence spread, obviously sick passengers or crew would also likely have been denied passage. An intervening stop or series of coast-hugging short hops would have provided no guarantee of an infection-enhancing urban environment from which intervening infections could have erupted. Passengers who just began to show symptoms near to arrival would not alter epidemiological modeling greatly except potentially to increase the chances of infection spreading more rapidly ashore post arrival. Those reaching the more debilitating and potentially terrifying later stages (see below) of the progression of MARV lineage pathology might well have been isolated or even subjected to extreme prophylactic measures by superstitious and terrified crews (such as disposal overboard)—perhaps even regardless of social status. Dead bodies would have been also disposed of in such an unceremonious manner. While this would have been even more relevant with an aerosolized pathogen, there are other modeled factors that also would have limited MARV lineage transmission in this specific spatial maritime epidemiological environment. Isolation of infected individuals would have, whether out of fear or the application of the simple disease avoidance strategies practiced by Romans, been easier above decks. Sunlight would have played an, albeit limited, disinfecting role. And to apply modern analogues, those aboard would simply have been more “socially distanced” above deck than below (or for that matter in the crowded urban *insulae* characteristic of imperial cities). The ocean would have also provided an easy avenue to avoid contact with or quickly dispose of contaminated “medical waste.” Under these overall circumstances, relatively few epidemiologically viable new infections would have likely occurred while at sea, and, most importantly, longer voyages beyond the infection window/Tg of MARV would actually have served as something of an effective prophylactic quarantine. Those who deboarded with would have thus likely contracted it pre-voyage. This overall analytical conclusion constitutes a key factor in the way that our epidemiological modeling interprets the spread of the Plague of Cyprian.

Guardrail Modeling–Broader Epidemiological Contexts

Part III, overall, thus constitutes a plausibility/probability test for our Plague of Cyprian solution. Does a coherent and more refined inferential picture, supported by evidence, of the epidemiology and pathology of the Plague of Cyprian emerge or not? Do our conclusions fit both the available ancient *and* modern evidence? Most crucially, this complex overlay must explain why a third-century outbreak of the Marburg virus or one of its ancestors would look so different from a modern Marburg outbreak? Or to put it a different way, do the evidentiary pieces of the puzzle concerning this ancient horror seem to fit—in terms of matching our proposed outbreak modeling to what we know about the Plague of Cyprian from the available ancient evidence detailed in Part I? Can additional modern scientific evidence be incorporated and further conclusions drawn that result in a more fully consilient and complete picture of the Plague of Cyprian?

In this undertaking, we must once again be mindful of several specific guiding principles elucidated by Kyle Harper, both in *The Fate of Rome*'s broader examination of the phenomenon of ancient pandemics (particularly in relation to his treatment of the second century Antonine Plague) and later in his 2021 chapter essay "Germs and Empire" in *Empire and Religion in the Roman World*. While Romans could greatly fear pestilence, as they undoubtedly did the Plague of Cyprian at its various peaks, they were also somewhat used to a prevailing high level of infectious disease. Roman cities, in particular, were annual demographic sinks of mortality due to contamination and illness—particularly in the summers with outbreaks of malaria and various gastrointestinal illnesses. Roman superstition, medical humourism, *topoi*, or political propaganda could also create distorted perceptions of disease. Harper suspects, for example, that Roman attributions of the origin of the Antonine Plague to the Parthian campaign may have been epidemiologically wrong and "colored by imperial propaganda regarding [the failed usurper] Avidius Cassius," while the real origins of that pandemic event may have lain in the maritime Indian trade being conducted through Red Sea ports [21].

In whatever manner Romans explained it, a "welter of infectious disease was always present" and indeed expected. Early manifestations of a plague would probably have elicited little note. Cyprian's passing mentions of plague during the months before the fall of 251 CE (...and "On Mortality"), detailed in Part I, can be interpreted in this vein. Only once the disease had built to truly epidemic levels, would official and popular attention have been truly engaged. While the "large-scale medical apparatus based on home visits" of the Roman Empire did provide some effective palliative and other treatments and quality basic nursing care in many instances (potentially significantly cutting into fatality rates from disease), its practices were based on humourist ignorance of the true causes and mechanics of disease transmission. Romans had no concept of how infectious microbes actually "circulated... disease throughout the cities." However, a "rough and ready" logical awareness of the communicability of disease also existed—particularly among the more educated elements of Roman society. Another rather happenstance defense "against the diffusion of a pathogen" was the "friction" created "by the time and expense of travel" in the ancient Roman world. One could add, to Harper's broader highly informative maxims, an additional note about the often rather physical nature of overland travel—particularly when dealing with a disease as debilitating as MARV. This limitation would apply even when utilizing the Empire's impressive road systems. Due to these reminders, Harper insightfully cautions that analysts of ancient pandemics should eschew simple analogies of "ever-widening concentric circles" but rather see each plague event "as a toxic and fissile pinball.... borne by [patterns of] human movements" [20, pp. 103–108]. This convincing conceptual approach of Harper's greatly informed our following Plague of Cyprian modeling and epidemiological analysis.

In adapting Harper's highly worthwhile concepts for more specific general guiding or, in our terminology guardrail models, it is important to keep in mind that retrospective diagnosis is a necessity. Realities, whether factually confirmed or analytically projected here, are complex. Plugging in the epidemiological biologies and pathologies of differing pathogens would completely change the resulting models. Aerosolized pathogens like smallpox or measles would produce very different modeled epidemiological-pathology environments and projections than a MARV lineage pathogen despite the fact that pathogen biology is but one component of a disease's basic reproduction number—R0. We believe that we have made a strong case for a MARV lineage pathogen in Parts I and II, however, and we will now proceed on that basis. Imperial road systems, for instance, were, of course, extensive, covering over 250,000 miles [17, p. 9]. But a MARV lineage pathogen would not have been easily dispersed over them in an epidemiological sense. A combination of time, distance, relative lack of pathogen infectivity, effort, debilitation from applicable pathology, sunlight, potential abandonment (of either the journey itself or the infected person(s) on the way), often low population densities between urban nodes, etc. would have made this extremely unlikely, in our view, beyond the basic MARV infection window of up to ten days.

As we traced the pestilence's likely maritime entry pathway from Adulis to either Berenice, Myos Hormos or both (probably in the fall of 248 CE) in Part II, it is thus necessary to examine how the virus might have spread from one or both of those places. Both ports required an overland journey to Luxor/Thebes (very notable in light of the archaeological findings there detailed in Part I) and very nearby Coptos, where fall barge travel with the south to north flow up the Nile would bring a traveler or merchant to Alexandria in around 8 days; the latter with relatively little further physical effort. According to Dionysius/Eusebius, the Nile was also flowing again—making such a less arduous and quicker journey eminently possible. The reverse direction against the current would have taken longer—around 12–13 days. The well-appointed Via *Hadriana* road connected Berenice to Luxor and Coptos, but the journey involved 285 modern miles and consisted of around 11 days of travel; this is outside the average exposure to infection window for MARV (of 5–10 days). It thus marks our first potential instance of a potentially outbreak-limiting ancient travel-related epidemiological "friction." The ratio is close enough that it might not have been severely limiting but likely at least somewhat. Limited numbers and low effective levels of population density on the road and victims coming down with Marburg symptoms (and becoming debilitated) before reaching their destination, would have, even with the weakly enabling presence of Roman road stops/way stations, posed a very significant epidemiological impediment to the spread of such a debilitating non-aerosolized, and highly lethal pathogen. Such an epidemiological dissemination path would have been slow, sporadic/frequently interrupted, and low volume with lessened chances of generating a larger outbreak or more long-range dissemination. But by the third century, the port of Berenice had largely fallen out of favor (its fortunes would revive somewhat in the fourth century) and Myos Hormos (which was located near the modern Egyptian city of Al-Qusayr) had become the favored choice for the Aksum-Middle East-Africa-India maritime trade. From there, the overland journey of 4–5 days to Luxor and Coptos would have been well within the average Marburg infection window and far less of a potential impediment in regard to spreading to more densely populated urban centers. An, in context, highly potentially infectivity-enhancing practice of such journeys on this route was for likely heavily perspiring travelers to take shelter from the burning desert heat in the afternoon in cramped, dark caves. For a pathogen like MARV, this would have been quite an ideal epidemiological environment. Isolation of anyone suffering MARV symptoms in such a spatial environment would have been likely impossible; epidemiologically significant contacts could also have, under such circumstances, reached well beyond any caregivers. The epidemiologically downstream impact of such inbound infection enhancement on Luxor/Coptos could also have been quite significant [1, pp. 45–50].

Myos Hormos to Luxor/Coptos overland, and on to Alexandria via the Nile—this was the likely initial epidemiological path of the Plague of Cyprian within the Roman Empire. What would have been clearly epidemiologically overlapping timelines for the Myos Hormos and Luxor epidemics could also explain the "Ethiopian origin" perceptions of Romans in the narrower sense of that term—the source regions of the Nile. Luxor could, perhaps because of the clearly applicable *topos*, have been (wrongly in our view) perceived as the entry point for the pestilence (in line with Roman perceptions of disease-causing polluted water and miasmas)—the POC could clearly have been perceived to have come up the Nile. An additional potential maritime alternative, though we don't have enough information about it to examine it in detail in this context, would have been the *Amnis Trajanus*. This was a second-century canal that directly linked the Red Sea and the Nile. Once again, the inferential and evidentiary puzzle pieces fit rather seamlessly. Additionally, the plausible, albeit hopelessly conflated, outbreak/dissemination window of around September 248 CE detailed in Part II would have precisely coincided with peak directionality and annual trade/movement (largely maritime and riverine) regional patterns. Additionally, if the pathogen indeed arrived in Alexandria in October (just as cooler weather was beginning to prevail there) as seems likely, the plague's prevailing seasonality would have potentially accelerated the infection's spread in the city over the ensuing months [1, pp. 45–50].

There is also some remarkable congruence between the way Kyle Harper depicts the Plague of Cyprian as spreading in terms of its timeline, the temporal and seasonal patterns of travel within the Roman Empire, and the epidemiological profile of the MARV pathogen. From its deadly major urban center debut in Alexandria in the spring of 249 CE, the plague seems to have taken somewhere around three years to peak in Roman Africa's second city of Carthage—in the spring of 252 CE. And there being so memorably described by St. Cyprian. Yet, the timelines of Rome and Carthage (on different continents/separated by the Mediterranean) seem to have essentially been the same. This, at first glance, seems very puzzling. Additionally, the annual maritime closure, the *mare clausum* (November to February), would have been a factor in slowing the spread of any pathogen across the Mediterranean. But the overlay of the evidence of imperial travel and trade patterns considered within the application of Harper's conceptual schema of "friction" and "fissile pinballs" clarifies the picture. Road travel between Alexandria and Carthage took about 94 days. This would have made any spread of a non-aerosolized pathogen of Marburg's epidemiological profile extremely unlikely to be speedy and probably, for the reasons noted above, would have prevented it altogether through this overland pathway. Such road-system spread, beyond a fairly short range, would have required multiple radiating/cascading smaller "on the way" contact/infection hubs to "slow burn and build"—quite probably *in sequence*. The non-aerosolized "fissile" chain-reaction infection could easily have fragmented to the point of "burning out," in an epidemiological sense, from the weakening frictions of time, victim debilitation, expense, travel distance, and exertion overland. Similarly, seasonal interruptions or infrequency, expense or difficulty of maritime travel, and relative lack of epidemiological fuel (lower population densities) along the way between the two densely populated urban centers of Roman Africa would also have been limiting factors [35, 39, pp. 399–400].

A" coast-hugging" maritime journey from Carthage to Alexandria in the fall took around 13 days, but the unfavorable prevailing winds meant that the Alexandria to Carthage journey at the same time of year took at least 18 days (with lots of rowing potentially involved). Differences in

Roman military and civilian maritime technology must also be factored in. Roman merchant ships' v-shaped hulls meant that they could not operate efficiently as close to the coasts as Roman warships could. Such an Alexandria to Carthage voyage may well have, in fact, consisted of a series of "short hops" and potentially taken even longer and been costlier and less frequent (though also allowing for somewhat less physically arduous, somewhat more efficient, and speedier viral dissemination to intervening multiple contact points than with overland travel). Similarly, voyages from Alexandria directly to Rome would have taken at least 18 days in the fall as the prevailing winds were similarly unfavorable. Some historians even believe that such a direct journey would have been essentially impossible or effectively hopelessly cost-inefficient (potentially taking as much as 53–73 days in summer) at certain times of the year even outside of the (not absolute) *mare clausum* [6, 7, 35, 41, 42].

All these travel windows are also beyond the average exposure to infection window or the average Tg for MARV, and thus introduce considerable further elements of "friction" to our overall proposed epidemiological dissemination model. The coastal maritime journey from Alexandria to Carthage would have been against the prevailing winds. It seems a difficult dissemination pathway (either directly or through multiple radiating contact point intervening infection outbreaks), but any presumed spread would have been quite slow, effectively inefficient, potentially fragile in its sustainability, and incremental. The "rough and ready" Roman caution regarding communicability as the disease slowly built and spread through such "fractal" and gradual means would also have had more time to develop and be applied; such modified behavior might well have further slowed or stopped the pestilence. Conversely, the prevailing winds were much less adverse between Carthage and Rome; fall maritime journeys took only 2 to 4–5 days [6, 7, 35, 42].

Rome was also a massive consumer of foodstuffs and the region around Carthage was a leading producer. The volume of trade and travel in that direction would have been also structurally a better economic fit, and thus generally higher than between the two Roman African agricultural-exporter metropoles. The essentially shared timeline of Carthage and Rome thus makes as much sense holistically as the highly divergent one between Carthage and Alexandria. Very many (if not literally all) roads and sea lanes did, proverbially and actually, "lead to [and from] Rome", but some took much longer and were less traveled and/or more expensive than others. From Rome, however, the central hub of the mid-third century Empire, the Plague of Cyprian would have sent careening "toxic pinballs" in a multitude of directions very widely both shorter range overland and longer range via the cheaper and faster imperial "maritime highways"—the Mediterranean and the empire's network of European riverine arteries. Port cities such as Alexandria, Carthage, as well as Athens and other Greek port cities, were apparently, and notably, hit very hard by the Plague of Cyprian—as were towns on major riverine arteries. A more likely (albeit highly speculative) path, in our modeling, would have been originally eastward to Antioch (from Alexandria), west to Perge, Athens, Tarentum, and finally to Rome. This dissemination pathway would have, essentially through the limitation of the full effects of the unfavorable prevailing winds, kept all the travel times between urban nodes within the requisite MARV infection window. Yet each urban node would also have had its own complex epidemiological dynamics and its effects upon surrounding towns and regions; there are also seasonal variations and the effects of the *mare clausum* to consider. Any sort of precise calculation is thus impossible. A plausible projection seems to be the analytical/modeling limit in this instance, even in a guardrail sense. Regardless of the exact path, epidemiological dissemination would have taken considerable time and probably would have been an overlapping and occasionally interrupted process. Working, even tangentially against the prevailing winds and the annual *mare clausum,* the disease could indeed have taken years to reach Rome. We believe it took around three—from initial dissemination into the Empire during the fall of 248 CE (as noted in Part II) to the fall of 251 CE and the beginning of the pestilential-defined two-year reign of the emperors Gallus and Volusianus. Indeed, the early 250 s CE seems to have been the height of the first epidemiological wave of the Plague of Cyprian. Gallus and Volusianus are described by the Christian writer Orosius as "being famous for this plague alone," a "pestilence of incredible diseases [that] extended" to nearly the entire empire. The pagan historian Zosimus comments that the plague ravaged the Roman Empire to such an extent that "nor was so great a mortality ever known in any former period" [20, pp. 138–140; 36, p. 315 (7.21); 52, p. 8 (1.16)].

Even considering the possibility for considerable hyperbole, this evidence regarding the timing of the plague's peak works very well with the model of it dispersing outward from Rome as a central node in mid to late 251 CE. The sea lanes and riverine arteries would have been major and relatively rapid pathways of dispersion. Overland dissemination would have been shorter range and more uncertain, and many "toxic pinballs" would have fizzled out along the way. Some localized non-urban dispersion would also have been aided by the panicked exodus from plague-ravaged cities and towns—as the evidence (noted in Part I) of Dionysius/Eusebius documents for Alexandria and of Cyprian and Pontius for Carthage. The evidentiary overlay of modern filovirus-related *ville morte* events, also noted in Part I, similarly applies. All of this more specific evidence and analysis aligns well with Harper's initial Plague of Cyprian epidemiological timeline and guiding concepts regarding ancient pandemic dissemination patterns. The holistic picture also coherently fits with the overall epidemiological profile of MARV. Having examined the Plague of Cyprian's potential broader pattern of dissemination, the time has come to examine the inferred epidemiological mechanics of urban infection. Modern epidemiologists, such as those of the Institute for Health Metrics and Evaluation at the University of Washington (who have been a leading source of modeling/forecasting for the Covid-19 pandemic in the United States), model both historical and modern disease outbreaks based on a calculation/concept known as R0 in their forecasts. R0, or basic reproduction number, as described in the earlier parts of this book, is indeed a much-misunderstood concept.

Grounded in more limited divisions of a disease-affected population into groups who are "'susceptible', 'infected,' and 'recovered'" that date back to the 1920s, R0 is a scientific concept that originated in the 1950s. At its simplest and narrowest, it represents the "average number of people infected by one infectious individual" carrying a specific pathogen. Or to put it another way, "the expected number of secondary cases." A number of R0-based models, or significant components of them, have repeatedly proved to be strikingly accurate. Yet, this foundational predictive number is highly variable over a range of different scenarios, pathogens, and environments. Recent outbreaks of coronavirus disease in the twenty-first century, SARS, MERS [Middle Eastern Respiratory Syndrome], and now again with Covid-19) have also demonstrated the "notoriously tricky" difficulties involved in nailing down precise R0 values and using them as the foundations of mathematical calculations designed to precisely model and predict the epidemiological progression of disease outbreaks. The concept also is "often misconstrued [by non-specialists] as a fixed property of a [given] pathogen." R0 values actually vary widely for the same pathogen due to differing "risk factors...among outbreaks." Yet confoundingly, basic biological infectiousness *is* also indeed a major contributing factor. Other factors include patterns, types, and frequency of human contacts and movements. Climate, societal patterns, behaviors and settings (urban/rural), population densities; ease, cost, and speed of movement, seasonality (and the human social response to it), and geography can also all affect R0 values as does basic pathogen biology. While it is possible to consider the latter concept of a baseline "biologic" R0, in a limited comparative sense at least for specific pathogens, it is also important to note that such a calculation would have little actual real predictive value separated from the context of a more complete set of variables. Thus, due to the complicated nature of R0, the measure is often, "misrepresented, misinterpreted, and misapplied." Some epidemiologists indeed recommend using a second modified R0 -type measure called Re (the "effective" reproductive number; or sometimes Rt, its variation over time). Re-based mathematical modeling, in a modern setting, allegedly also accounts more fully for changes in social and individual behaviors (such as social distancing, lockdowns, or mask wearing in the context of the Spanish flu or Covid-19 pandemics), development of immunity in affected populations, or for more direct prophylactic medical or organizational interventions (such as full-scale quarantines or vaccinations). Other epidemiological modeling experts attempt to incorporate such components into a recalculated R0 value [11, 28, 51].

R0 and Re/Rt values are expressed mathematically and, if the disease outbreak is to be basically sustained, must always exceed 1.0. R values are usually expressed as a range. Some recent examples include Covid-19 in 2020 (with an R0 range of 1.4–5.7), MERS in 2014 (0.45–3.4), the West African Ebola outbreak of 2014 (1.5–2.5), SARS in Hong Kong in 2003 (1.7–3.6). The historical range measure for the great Spanish flu pandemic in 1918–1920 was 2.2–2.9. Twentieth-century outbreaks of measles demonstrated a yet far higher degree of infectiousness (12.0–18.0). Attempts (potentially invalid in the eyes of some experts) have also been made to determine a more fixed infectiousness value for present or historical outbreaks in a more "generic" environment. Calculations done with this goal in mind, of course, seek to bypass some of the inherent variability of R0 and Re/Rt values. Some of the comparative scores produced are values of 1.2 for common influenza, under 1.0 for MERS (meaning that many of its outbreaks are inherently not sustainable), 2.0 for Ebola, 7.0 for smallpox and 15.0 for highly infectious measles [16, 51].

All of this produces a shifting and highly confusing picture. While R0-based calculations are often strikingly epidemiologically accurate, other such calculations by leading experts in the field have proven to be wildly inaccurate. The CDC, for example, produced an estimate of 500,000 Ebola infections for a geographically wide sector that was part of the largest-ever recorded Ebola outbreak in western Africa in 2014. This turned out to be an overestimate (actual infections totaled only about 25,000) by around a factor of 20x. The difficulty seemed to stem, in retrospect (as the specialists involved examined their own work), from the mistaken generalization of assumptions concerning certain local behaviors and social practices at the outbreak's epicenter to the broader region across the entirety of the nations of Liberia and Sierra Leone. Similarly, the range of R0 calculations by several teams of experts for a single measles outbreak produced values from 3.7 to 203.3! Problems of interdisciplinary consilience also bedevil such calculations; should R0 or Re/Rt ranges be calculated by mathematicians, sociologists, virologists, or interdisciplinary teams composed of all three types of specialists? Or, if examining a historical pandemic, should historians or even archaeologists be included? All of this makes R ranges fluid, uncertain, and hard to calculate as they attempt to replicate an interconnected multitude of human and environmental factors in terms of interaction with a given pathogen. The resulting infectiousness values are thus not fixed but rather dependent and transient in their essence. In the words of Dr. Paul Delamater of the University of North Carolina, "R0 values are nearly always estimated from mathematical models, and the values are dependent on numerous decisions made in the modeling process. R values are thus not intrinsic fixed qualities of a pathogenic microbe, but rather encompass all the complex realities of a pathogen's interaction with the intricacies of climatic, biomedical, historical, physical, geographic and social environments." The same pathogen, in differing circumstances in terms of time, population, social behavior, settings, and place, might thus demonstrate very similar or wildly varying R0/Re values. Thus to return us to our ancient Roman context, overall, we somewhat *disagree* with the first part of Kyle Harper's statement in *The Fate of Rome*, regarding the second century Antonine Plague, that, "Pathogen biology determines the dynamics and dimensions of a disease event [it isn't actually that straightforward]", but *profoundly agree* with the second part of that same statement that, "…if we know the identity of the microbe behind the Antonine Plague, we can hope to fill in certain missing pieces of the puzzle." Additionally, in our context, such analysis, despite the inherent, and here freely acknowledged, risks of circular and reductionist reasoning, can also provide additional (and mutually supporting) strong evidence for the original identification of the pathogen through the creation of workable evidence-supported modeling that achieves consilience with the overlay of *both* the existing applicable ancient evidence *and* modern scientific medical knowledge. Given the grave inherent challenges and uncertainties, however, we will approach this task cautiously and generally rather than boldly and specifically. In the highly applicable words of Dr. Jeffrey Shaman, Director of the Climate and Public Health Program at the University of Columbia's Mailman School of Public Health, "Models are like 'guardrails' to give [merely] *some sense* of what the future *might* hold [emphases ours]." Except, in our context, it is not the future but rather the distant past of around 1,772 years ago that we are modeling. Having laid out our analytical course, we can now begin to examine the projected epidemiology of the Plague of Cyprian [11, 49, 51, 20, p. 102; 28].

Given the complexities and the very incomplete and non-specific nature of these applicable historical data, it would be, in our opinion, impossible to try to calculate an exact R value for the Plague of Cyprian with anything approaching precision. We can, however, effectively use our ancient–modern evidentiary overlay, together with the core components embodied within R0 and Re/Rt values, to examine more broadly, and yet also in a surprising amount of detail, how a third-century Roman outbreak of a Marburg lineage virus might have been fundamentally similar to, but also have probably diverged vastly from, the patterns of twentieth and twenty-first century MARV outbreaks. To put it more simply, how would the pathogen have specifically interacted with what is known about the society and structures of the Roman Empire of 248–262 CE? Given that modern medical prophylactic measures obviously wouldn't have been in place in the imperial third century, and given that the available evidence indicates that the Plague of Cyprian was, in the third-century context, a novel disease, we have generally if not exclusively expressed R values in terms of R0 in the following sections. We must start with a more detailed re-examination of the basic properties of the Marburg pathogen itself. While we have covered MARV in Parts I-II, there are certain characteristics that are most important to elaborate upon and review at this point. MARV strains are single-strand RNA viruses composed of seven different proteins. MARV structures are basically similar to but also display differences from those of the closely related Ebola viruses. As noted earlier, the exact specifics of zoonotic transmission are not known, but the Egyptian rousette bat is the natural reservoir of

MARV and the zoonotic source for human infections. The natural reservoir of EBOV is also suspected (but not yet proven) to be an unspecified species of bat. One suspicion is that in the dark environments inhabited by *Rousettus* such as Kitum Cave in Kenya, the Kitaka mine and Python Cave in Uganda, or the illegal gold mine that was the center of the outbreak in Durba-Watsa in the DRC, MARV infections occur through contact with bat guano—either directly or with prolonged exposure in the guano-particle saturated, caverniculous environment. This involves the inhalation of a semi-aerosolized powdered particulate form. Notably, MARV is considered a non-aerosolized pathogen with a few such caveats. The "risk of aerosolized transmission of Marburg virus (MARV) in the natural setting is believed to be low." "Prolonged exposure" to mines or caves inhabited by *Rousettus* bat colonies is, however, the typical zoonotic infection pathway in humans; this reality holds true particularly in mines [2, 8, 13, 25, 31, 34, 47–49].

The virus is, however, stable in aerosols and nonhuman primate studies have demonstrated that MARV is "utterly" lethal following experimental aerosol exposure. This has led to twenty-first-century concerns about bioterrorism. Both Marburg and Ebola have also shown, at least in primate transmission, some highly limited more "natural" aerosolization under specific circumstances. While its outbreaks display a fairly wide range in terms of lethality, the average overall fatality rate for MARV human infections in outbreaks from 1967 through 2017 is statistically around 80% (and as high as 88% in some outbreaks). Such high "case fatality rates" mark MARV as one of the most lethal human pathogens known to medical science. Some experts argue for a lower "average" CFR of 50%-65% assuming "strong and modern case management." As of 2022, there is still no effective vaccine for Marburg though "blood products, immune therapies and drug therapies are under development." Good palliative nursing practices, such as proper hydration, administration of potassium supplements, antiemetics (this last modern treatment particularly notable in light of countervailing common Greco-Roman medical practices), administering oxygen and other modern symptomatic treatments somewhat "improve survival rates" in an advanced modern clinical setting. As noted earlier, the incubation stage can be as brief as 2 days and as long as 21 (with the average incubation or infection window period of 5 to 10 days). Human-to-human transmission is particularly likely if it occurs percutaneously (through the skin) or muco-percutaneously (through the mucous membrane). The pathogen can sometimes persist in the eyes, in the testicles and semen of men, and in pregnant or recently pregnant women—in the placenta, amniotic fluid, and breast milk for a number of months. Male survivors should practice safe sex for up to 12 months or until their semen has tested negative twice. Uncommon asymptomatic replicative infections for the closely related EBOV were documented, however, in medical literature in the early 2000s. Under prevailing modern WHO protocols, those eligible for contact tracing include persons who "slept in the same household", had physical contact with a living or dead infected [or suspected of being infected] person or his or her feces or bodily fluids, or their clothes or bedding. Breastfeeding, medical setting contact, and lab exposure to the pathogen are particular causes of concern [8, 31, 45, 47–49].

Modern medical prevention/education and response teams, trying to prevent MARV outbreaks or rapidly end them, are faced with a number of challenges. They must ensure that all logistical challenges, such as adequate supplies of PPE (Personal Protective Equipment) for field and medical settings, are met. Ideally, they should be effectively employing modern standard infection control protocols/epidemiological investigation techniques—such as contact tracing, active case finding, with referrals to care units, monitoring of recovering patients, conducting social awareness/mobilization campaigns that encourage both pathogen-specific and general medically hygienic practices (such as handwashing), identifying infection sources, establishing field laboratories and clinics and undertaking rapid and valid sample collection as well as carrying out accurate laboratory testing. Additionally, potential key, and particular to MARV, R0 amplifying factors must be accounted for and mitigated. One of these is potential contact between Egyptian rousette bats and pigs, the only domesticated animal known to be able to contract and shed MARV. Pigs thus can plausibly function as "potential amplifier hosts" and activity at pig farms must be monitored closely. Another is education specifically aimed at miners designed to mitigate environmentally elevated zoonotic risks. Others relate to human-to-human transmission prevention efforts. Specialists, often cultural or medical anthropologists, should note and "pay attention to social and cultural practices and events that could transmit and amplify infection" [47–49].

Extra attention should also be devoted by medical specialists to MARV R0 -amplifying epidemiological environments such as those associated with medical care. "Special risks to medical staff due to the ease of transmission [in such environments] ...can have a profound impact upon the communities they affect." Infection control precautions in the health care setting (whether that be a hospital, field clinic, or home care setting) should be highly exacting and rigorously enforced. The latter environment should be avoided if at all possible (though cultural considerations might make that impossible). Regardless, "strict compliance with biosafety guidelines" should be observed. Patients should ideally be housed in single isolation rooms. If this proves impossible, patients should be cohorted with laboratory-confirmed and merely suspected cases separated. It is also very important that each patient, regardless of health care setting, must use a separate "dedicated" toilet or latrine. Barrier nursing (including use of PPE) procedures, avoiding touching the patient (whether living or dead) or their fluids without protection under any circumstances, disinfection or destruction of contaminated objects, limiting physical contact with infected persons even when wearing PPE and otherwise maintaining a distance of 1 meter (if at all possible, behind or to the side of the patient when in proximity) even when wearing protective gear are essential prophylactic measures. One meter should also be the distance between beds if it is necessary to house multiple confirmed infectious patients in a single unit. Visitors (lacking PPE) should generally be prohibited, but if this proves destabilizing to the overall community care environment, alternatively kept at a distance of no less than 3 meters [25, 47].

Extra "infection control" measures should be taken in regard to the highly contagious blood and other bodily fluids and any potentially contaminated surfaces. Contaminated linen either should not be reused or instead destroyed (preferable) or at least not washed by hand. All medical waste should be disposed of under strict protocols and preferably safely incinerated. Certain medical procedures, such as phlebotomy should be banned in most instances. In fact, severe limits should be placed on even the presence of syringes, needles, scalpels, and other sharp objects capable of causing percutaneous (mucocutaneous transmission is also a highly dangerous possibility) injury to caregivers. Such objects should also, if used, ideally never be reused ("rigorous" use of dedicated equipment for each patient should be the standard) or, if so, especially thoroughly decontaminated. We will contrast these specific modern practices to their analogues in the Greco-Roman medical setting presently. These modern cautions also apply to BSL-4 (Biosafety Level Four) labs; laboratories below that top certification level are not permitted to handle

MARV. Non-medical personnel such as security, cleaners, and housekeepers should be kept away from the unit in which MARV patients are housed. Proper PPE in the unit should include, at minimum, double gloves, a medical gown or coverall, an apron, a face mask, supplemental eye protection such as goggles or a face shield, a hood, and covered boots. Adherence to such protocols is literally a matter of life and death. "Health care workers have been frequently infected while treating patients with suspected or confirmed MVD (Marburg Virus Disease). This has occurred through close contact with patients when infection control precautions are not strictly practiced." Transmission in such medical settings [particularly from a needle stick or other percutaneous injury] is also "associated with more severe disease, rapid deterioration, and possibly, a higher fatality rate." [48, 49].

Funerary practices are another modern key potential R0 enhancer. "Burial ceremonies that involve direct contact with the body of the deceased can also contribute to the transmission of Marburg"—particularly among relatives and friends of the victim. In much of the MARV-endemic region, funerals in which many participants touch the body, can last for days. Yet, "prompt" burials carried out in "a safe and dignified" manner are the relevant epidemiological control benchmark of twenty-first-century rapid response teams in outbreak zones. Under no circumstances should burial be delayed or remains "washed or embalmed." Outbreak response teams, often composed of cultural outsiders, must walk a fine line and be sensitive to the potential for counterproductively angering or terrifying the affected community by disposing of the bodies of MARV or Ebola victims "without consideration for customary funeral rites." At times, medical NGOs have resorted to the deployment of compromise "ethnographically-based videos on safe funeral rites." These are produced in order to simultaneously attempt to promote epidemiologically safer funerary rites and reassure affected communities. Another source of potential cultural conflict in African filovirus outbreak zones are the practices, often conducted in harder to monitor in-home care settings, of traditional (folk) medicine practitioners or religion-based faith healers. Their R0-amplifying practices, in context, include the use of multipurpose or reused equipment, injections, failure to follow strict disinfection protocols, scarification, laying on of hands, as well as the use of purgatives and emetics [47–49]. As we will see, third-century Roman urban settings were characterized by a large number of these very sorts of modern "red-flagged" practices or their close epidemiological equivalents.

As generally non-aerosolized pathogens, the members of the deadly African *filoviridae* are (very fortunately for humanity) not particularly contagious under most circumstances. The European Center for Disease Prevention and Control commented, "Therefore the risk of [EBOV and MARV] infection is considered very low *if appropriate infection prevention and control precautions are strictly followed.*" Transmission is, as noted above, generally through "mucosal surfaces or breaks or abrasions in the skin" via contact with infected animal tissue or through human vomit; saliva, sweat, feces, breast milk, urine, tears, and other fluids or discharge from the eyes and blood. As noted, infected corpses are extremely dangerous, and the virus can also be transmitted up to 12 months later through fluids associated with sexual contact. Typically, sufferers during the pre-symptomatic incubation period are not believed to be able to transmit MARV, but this is not an entirely settled question. Because of the need for close contact transmission, and the fact that even some seemingly strong exposures often don't end in infection, many Marburg and Ebola outbreaks flame out quickly or are rapidly contained, after detection through ELISA or PCR testing of possible victims, which triggers modern medical intervention. Marburg may also be less contagious than Ebola as it has not produced as large outbreaks (hundreds vs tens of thousands of dead); though this discrepancy may also be simply due to MARV's less frequent and regular (sporadic vs cyclic for Ebola) outbreak appearances. The ability to sustain large outbreaks in the last 25 years has, however, shown MARV to be "a serious public health threat." As noted, unlike EBOV, there are also no existing vaccines for MARV [13, 31, 16, 8, 29, 39, pp. 45–130 and 263].

The R0 of EBOV outbreaks has often been calculated in the 1.5 to 2.7 range. MARV, as it has sustained large outbreaks, probably also has a similar but possibly slightly lower average comparative R0 range. This possibly slightly lesser level of infectiousness and slower disease progression is probably due to the less efficient binding/replicative properties of some MARV proteins in comparison to the homologous structures of EBOV. Despite Ebola's far greater outbreak frequency, regularity, and higher overall death tolls in recent decades, some experts contend that due to the similar levels of extreme lethality, slower disease progression, and the more prolonged average death times for MARV, "Marburg deserves as much attention as Ebola virus when considering highly pathogenic ...VHF agents." Both MARV and EBOV were listed in the 2018 World Health Organization's Priority Disease List. The list was composed of diseases that should be particularly prioritized in terms of "research, development of countermeasures and emergency preparedness response." Particular *foci* for infection for both members of the African *filoviridae* have included interaction with infected animals, human-to-human community spread in hospital (often through direct contact, insufficiently sterilized needles, or other medical equipment, bedding, and clothing) or other medical/caregiver settings such as home care or "laying on of hands-style faith healing," contact with the corpses of victims (particularly when handled in complex funerary practices). Both of these latter types of infection are even more lethal in terms of the Marburg virus than other MARV infection pathways. Caregivers in hospital settings are indeed at heightened risk of contracting both members of the African *filoviridae* unless all applicable modern prophylactic measures are carefully followed. Both of the African filoviruses have also, due to infection through contaminated blood and also perhaps because of the very high lethality of infected corpses, shown an association with human violence and disruption, which also serve to disrupt medical interventions [31, 34, 16, 49, 29, 5, 39, p. 143; 8, 30, 44].

Marburg disease progression involves three stages: stage 1- days 0–5, generalization phase; stage 2- days 6–13, early organ phase; stage 3- days 14–21, late organ or convalescent phase (many survivors do not reach Stage 3). The times here represent slower evolutions of pathology. In many cases, the disease moves much more rapidly with deaths occurring between 5 to 9 days after symptoms first commence. As the sufferer, at whatever pace, progresses through the stages, stage 1 flu-like symptoms prevail at the outset (headache, diarrhea, fever and sweating, alternately chills, sore throat, myalgia, etc.). These pathologies will become more severe through stage 2 and result in fatigue, more extreme diarrhea, and vomiting. Bloody conjunctivitis, esophageal lesions, extreme prostration, bleeding sinuses, and general vascular permeability have also been noted at this stage. The often-fatal stage 3 symptoms include liver, kidney, or other organ failures, delirium, bloody diarrhea, blood from the eyes, dehydration, and other, at times terrifying, hemorrhagic manifestations. As noted, MARV is particularly persistent in the eyes and testes of survivors. This phenomenon has been duplicated experimentally using macaques. While survivors do develop antibodies and apparent immunity to the disease, they also often suffer a wide range of potential aftereffects including hepatitis, *sequelae* involving damaged joints, limbs, or organs, weakness and confusion/delirium, and a range of sensory—particularly ocular problems. There is indeed a pattern of "severe neurological consequences in survivors of filovirus disease." The duration of immunity in filovirus survivors is uncertain and some research indicates that immunity to MARV may be somewhat less durable than immunity to certain EBOV species; though, once again, this is not certain. "By their very nature, Ebola and Marburg, highly fatal and painful illnesses with 'terrifying symptoms', cause fear and even panic." Flight, including *ville morte* events in afflicted villages and desertion of hospitals and clinics by health care workers, fearful "of a misfortune they struggle to understand" is a common response. Similarly, a variety of types of violence and instability, including private score-settling, generational and ethnic conflict, belief that the disease is caused by curses and witchcraft, etc. often accompany outbreaks. These sorts of superstitions, uninformed and fear-driven behaviors too would have almost certainly found very strong third-century Roman analogues. In the twenty-first century, culturally sensitive public relations and interaction, guided by anthropologists, is considered modern medical rapid response "best practice" to avoid counterproductive "further traumatization" of affected populations as "community engagement is key to successfully containing outbreaks" [29, 31, 34, 47–49].

Table below by Dr. Mark Orsag/ Dr. Amanda McKinney. While this table was created for MARV specifically, there would also be some overlap in pathologies with other VHFs.

The ancient sources cited in our Table 7.1 include both direct descriptions and also a few cases of implied notation of pathology. Table 7.1 also constitutes a large part of our detailed evidentiary foundation for List 2.1 in Part I. Overall congruence between MARV and the Plague of Cyprian in Table 7.1 is very, very strong. The limited areas where the overlap isn't complete are, rather simply, explicable. To further examine and explain how a Marburg lineage outbreak might well have functioned in ancient Roman imperial settings, we must, at least broadly, overlay the known transmission characteristics of MARV in the twenty-first century with what is known about the daily functioning of urban and rural settings of the third-century empire. While, as noted, the overall Plague of Cyprian's R0 value can't be calculated on the basis of the incomplete existing evidence, some of the components that would have influenced those values can be clearly determined and examined in a surprising amount of detail. We will first concentrate on urban settings, and then shift our focus to rural areas. There were a number of aspects of urban life that clearly would have affected the overall infection pattern of MARV or one of its lineage ancestors. These would, at least for urban areas, easily explain high R0 values (comparable to the largest MARV outbreak in 2004–2005 in Angola or potentially even higher) and larger outbreak scales and longer durations for the Plague of Cyprian than for modern MARV outbreaks. Even without taking into account the additional countervailing effects of modern medical intervention, this is clear. The urban areas of the empire would have been, we believe, epidemiologically very strongly R-amplifying environments for a MARV lineage virus.

While numerous aspects of life in third-century Roman cities seem conducive to the transmission/dissemination of MARV, two linked endemic institutions—the medical system and the baths—functioned in ways that would have made them, when analyzed through our ancient–modern systems-synthesis evidentiary overlay, effective "super exposers"

Table 7.1 Pathology of the Plague of Cyprian as described in the ancient sources/cross-referenced with the pathology of MARV

Plague of Cyprian/Marburg symptom or epidemiological feature	*Plague of Cyprian source(s) noted in…*	*Plague of Cyprian/Marburg Pathology/Epidemiology congruence*
"Ethiopia" Southeast Vector	Luxor Site. Dionysius/Eusebius. Porphyry, Gregory of Nyssa, Symeon the Logothete, George Kedrenos, John Zonares	Probably Yes[a]
Medical/Caregiver Transmission	Dionysius/Eusebius, Cyprian	Yes
Deadly highly infectious corpses, Caregiver deaths	Luxor Site. Dionysius/Eusebius,	Yes
Sequelae to limbs	Cyprian	Yes
Sequelae hepatitis	Not mentioned	Partial[b]
Sequelae sensory	Cyprian	Yes
Sequelae specifically ocular	Cyprian	Yes
Pain	Cyprian	Yes, inferred[j]
Weakness/Prostration	Cyprian	Yes
Bloody eyes	Cyprian	Yes

(continued)

Table 7.1 (continued)

Plague of Cyprian/Marburg symptom or epidemiological feature	*Plague of Cyprian source(s) noted in...*	*Plague of Cyprian/Marburg Pathology/Epidemiology congruence*
Maculopapuiar rash (sporadic)	Not mentioned	Partial[k]
Other hemorrhagic manifestations	*DeLaud*	Yes
Diarrhea	Cyprian	Yes
Bleeding sinuses	Not mentioned	Partial[c]
Esophageal lesions	Cyprian	Yes
Urban build to peak	Cyprian, Pontius, Dionysius/Eusebius, Orosius, Zosimus, Philostratus, *De Laud, Historia Augusta*	Yes[d]
Rural presence	Orosius, Zosimus	Yes
Highly pathogenic	Cyprian, Orosius, Philostratus, Dionysius/Eusebius, Pontius, *Historia Augusta*	Yes
Dehydration	Gregory of Nyssa	Yes
Delirium	Not mentioned	No[e]
Vomiting	Cyprian	Yes
Vascular permeability	Cyprian	Yes, strongly inferred[f]
Fever	Cyprian	Yes
Association with violence/displacement	*Historia Augusta*, /Dionysius/Eusebius, Zosimus	Yes
Cool weather peak	Cyprian, Dionysius/Eusebius, Symeon the Logothete	Yes
Indiscriminate/terrifying	Cyprian, Dionysius/Eusebius, Pontius, Jordanes, Orosius, Luxor Site, De *Laud,*	Yes
Unique	Pontius, De *Laud*	Yes
Follows drought	Cyprian, Dionysius/Eusebius	Yes
Bloody eyes	Cyprian	Yes
Clothing/Eyesight transmission	*Excerpta Salmasiana II*	Partial[g]
Lingering death/suffering/pain	Cyprian, Pontius	Yes[h]
Relentless	Cyprian. Dionysius/Eusebius, Pontius, Orosius, Jordanes, Luxor Site, De *Laud,*	Yes
Eyes generally central to pathology	Cyprian, *Excerpta Salmasiana II*	Yes
Potential extreme destructive effects of disease	*Epitome de Caesaribus, De Laud,* Cyprian, Pontius	Yes[i]

(continued)

Table 7.1 (continued)

Plague of Cyprian/Marburg symptom or epidemiological feature	*Plague of Cyprian source(s) noted in...*	*Plague of Cyprian/Marburg Pathology/Epidemiology congruence*
Necrosis organs	Not mentioned	Partial, would not have been noted due to cultural prohibitions vs human vivisection

[a]As indicated by multiple predictive models (noted in Part 3); confirmation by isolate or genetic testing of MARV presence in the Ethiopian Highlands Rousettus population would, in our opinion, elevate this to a full "Yes."

[b]With modern Marburg survival rates at around 20%, and this being a potentially lethal sequela in an ancient setting in particular, the absence of a description in the ancient sources of specifically-defining hepatitis symptoms such as jaundice (which do not medically present in all cases) is not surprising in the least.

[c]This symptom was very prominent in the 1998–2000 Durba-Watsa outbreak in the Democratic Republic of Congo, but not in other modern MARV outbreaks; so once again, its absence in the ancient Plague of Cyprian sources is not surprising.

[d]The 2004–2005 Uige, Angola MARV outbreak (the largest ever) occurred in an urban (generally medical) setting and killed hundreds of people. The massive urban outbreaks indicated by the ancient sources regarding the Plague of Cyprian require some additional analysis (summarized later in Table 7.2) to explain in terms of the apparently greater scale.

[e]Cyprian's writings contain several passage that may or may not be oblique references to the symptoms of confusion/delirium, but they did not, in our opinion, rise to a sufficient level of analytical probability to include as a positive or even partial finding.

[f]This symptom, in an ancient context, would have been extremely unlikely (due to insufficient understanding of the circulatory system) to have been directly noted even by a physician- much less someone like Cyprian. But the "diseased putrefaction" of limbs noted by Cyprian is a very strong indirect indicator.

[g]Transmission via clothing/bedding is completely consistent with MARV. Transmission by gaze is either a misapprehension that potentially indicates naturally-occurring aerosolization (which would not be consistent with MARV) or a misleading cultural/historical artefact. Harper convincingly leans toward the latter interpretation, and we concur. See Harper, *Fate*...pp. 140. 333/Note 45.

[h]Pontius' account here is open to interpretation or possibly slightly contradictory, but, on balance, we judged it supportive of this point. Cyprian's narrative is strongly supportive.

[i]The *Epitome* describes the co-Empenor Hostilian as being "consumed by the plague." The similarly anonymous *De Laud Martyri* describes the "bloody destruction of ravaging diseases", Cyprian the "horrible and deadly" disease's "devastation", and Pontius the "excessive destruction of a hateful disease" with the numerous dead being "no longer bodies, but the carcasses of many." Such devastating symptoms do not necessarily present in a majority of modern MARV cases, with the bleeding being largely internal. But they can... For a graphic modern account of such a destructively transformative and physically-devastating Stage 3 case of the Marburg virus, see Richard Preston's *The Hot Zone*, pp 17–24. Especially when considered independently, however, these ancient accounts could be alternatively explained as rhetorical/allegorical flourishes or descriptions of the mass putrefaction of urban Plague of Cyprian victims.

[j]Cyprian's language "a fire originated in the marrow" may well reflect joint pain, another common MARV pathology. Humourism attributed pain to imbalances of humours within the body. These imblances could cause, it was believed, the body to become overheated.

[k]Variable symptom, both in terms of outbreak and individual pathology. May or may not present.

for transmission of one of the deadliest pathogens on earth. While even many strong exposures might not have resulted in infection, the cumulative epidemiological effects of the cascading exposures from the R-based interactions of a MARV lineage pathogen with the functioning of these two omnipresent institutions of third-century-Roman imperial urban life would have been very likely devastating. A third very important R0- urban amplifier would have been prevailing third-century funerary customs. The findings in the following sections are summarized in Table 7.2.

Table by Mark Orsag/Amanda McKinney.

Roman cities, in general, were densely packed with crowds gathering for everything from court proceedings, daily life in the ubiquitous public forums or crowded *insulae*, poetry readings, games, chariot races, or the theater (some temporary theatrical structures in Rome could seat 80,000 patrons). By the third century, Roman funerary practices (discussed in more detail later) had shifted toward burial and more complicated week-long rites and away from quicker more simple cremation practices (a shift that would also have served as a very strong MARV lineage epidemiological enabler). Closely crowded urban centers featured shared food at quick snack counters. Human feces and urine were also regularly utilized by urban Roman tanners and urine by a city's fullers (the ancient Roman equivalent of dry cleaners). Both human waste products were essentially commodities; urine, in particular, was the subject of a thriving trade based on public urinals, and was, at times in imperial history, specifically taxed. Urine and feces were regularly collected from numerous public lavatories and sold to the fullers and tanners by a class of merchants specializing in the trade. At times, human and animal waste products (particularly from dogs) were mixed together. In our context, it is also important to note the potential for infected blood in human waste. As noted by the poet Catullus in the first century BCE, urine was also used by Romans to whiten teeth due to the high ammonia content in aged urine. While this is a highly suggestive epidemiological pathway when considering MARV, considerable caution is in order. Urine was generally aged for many Roman uses to maximize cleansing ammonia content. While the details of Roman practices are unknown, such processes generally thoroughly destroy viruses and bacteria. Considerable risk to collectors and early process handlers (whether free persons or slaves) can, however, can be inferred. Late process workers and end-product users probably faced little to no risk. Tannery workers, who seem to have been somewhat more

Table 7.2 Epidemiological risk factor modeling of a MARV lineage virus in a third-century Roman urban setting

Confirmed modern MARV transmission pathways	*Specific third century urban Roman environmental/epidemiological enhancers*
Zoonosis- *Rousettus* colonies	N/A
Zoonosis- Infected Primates/primate tissue	N/A
Contaminated medical equipment	Phlebotomy, surgery, medical practice generally, settings of war, disorder, violence, Roman military medicine
Skin abrasions	Mentagra and other skins disorders, Strigil Use, Shaving/Depilation, self-scraping (poor) practices in Baths
Vomit	Induced/Medical Use, caregivers[c]
Saliva	Widespread Prostitution/Sexual Contact Baths[d]
Sweat	Multiple Enhancers in Public Baths: Exercise in Palaestra[a] Massage, plus self-scraping at public baths by the poor, Dry Sweating in *laconium* and *sudatoria*[b] caregivers[c]
Feces	Tanners, Enemas/Medical Use, Communal Toilet Sponge in Baths and public toilets, caregivers[c]
Breast milk	N/A
Urine	Fullers, Teeth Cleaning, Communal Toilet Sponge in Baths and public toilets, Folk Medicine practices in rural and some urban settings
Tears	Shared Eye Salves
Blood	Phlebotomy, Etc/Medical Use, settings of war, disorder and violence, Roman military medicine, caregivers[c]
Semen	Widespread Prostitution/Sexual Contact Baths[d]
Highly infectious Corpses	Plague Urban "Build" Peaks, Settings of war, disorder, and violence, military settings, funeral customs, caregivers[c]

[a]Probably generally possible by some infectious individuals only in a very narrow window early during Stage 1. Though rare mild or asymptomatic cases could potentially have been an enabling factor as well.

[b]This was often recommended for the sick by Roman physicians.

[c]Family, friends, and household slaves in a civilian setting.

[d]Some Roman baths even had houses of prostitution directly attached.

likely (though this too isn't absolutely certain) to have utilized unadulterated human waste products in narrow aspects of their trade, were probably even more epidemiologically at risk than those employed in the *fullonica* [18–19; 26–27; 29, 37, pp. 288–299; 40, pp. 284–319; 50].

The nearly ubiquitous urban and military functioning of Roman medicine would, however, have been a more widespread and extremely effective epidemiological enabler for MARV exposure/transmission—particularly in a densely packed urban setting—a true "super-exposer"/infection-multiplier, ***in a sort of perversely reversed rendering of a modern Re value***. This is true both on its own and in conjunction with the second "super-exposer" institution that we will discuss later—the Roman baths. The R0 exposure enhancements involved here overall would likely have been mutually reinforcing and cumulative. The ancient–modern overlays between these practices and modern epidemiological "red flags" are refined and revelatory, surprisingly precise and quite striking in their analytical guardrail modeling outcomes. In fact, the baths and the medical system were also strongly, and with tragic irony in this instance, linked. Roman authorities were obsessed with maintaining good hygiene and sanitation in their cities. Dirty water was, correctly, seen as a main cause of epidemic disease. The system of aqueducts that maintained the water supply of the Empire's cities thus served, among many other crucial purposes, a medically prophylactic one. Roman medicine was shockingly advanced in certain ways: As in its ability to treat certain types of parasites, solidly grounded basic nursing skills and palliative care including hydration, the ability to perform complex surgeries, a standard of military medicine that would not be equaled in Europe for many centuries after the Western Roman Empire fell, the recognition of the efficacy of inducing vomiting to treat a variety of ailments, etc. Yet, Roman medicine was also profoundly ignorant of many disease basics and operated on the basis of fundamentally flawed overarching concepts. Furthermore, magic, superstition, misunderstanding, "miracles", and valid science were never fully separated within it. Strong moral prohibitions on dissections and vivisections also limited the profession's ability to evolve beyond its "quasi-rational" equilibrium. Roman doctors (actually often ethnic Greeks) were usually "freelancers" at liberty to "choose the places where they practiced." They, as a rule, sought more advantageous circumstances in urban centers and with the military; rural areas would have been (by far) the least frequented by the medical profession. There was individual and practicional overlap between military and civilian physicians, but the legions and the Roman navy also maintained their own complex and multi-tiered directly state-funded medical establishments. Archaeological evidence from Roman military hospitals such as those at the ancient garrison sites of Novae and Carnuntum, with their highly organized (even modern-seeming) spacing and isolation arrangements of sick bays, also efficiently confirms the effective "rough and ready" disease transmission awareness of Roman medicine referenced (from literary sources) by Kyle Harper [4, 20, p. 109].

While urban and military medical practitioners were divided among competing schools of thought—Empiricists, Rationalists, Dogmatists, Eclectics, Methodists (whose symptom-based approach often seems the most "scientific" from the modern perspective), etc., all were grounded to one degree or another in erroneous ancient Greco-Roman medical theories of the Four Bodily Humours—each associated with a corresponding element—Blood with Air, Yellow Bile with Fire, Black Bile with Earth and Phlegm with Water. Additionally, miasmas, vapors and noxious "winds were faulted [along with dirty water] for bringing insalubrious conditions" and disease. Magic, religion, and superstition were all intertwined with medical practice. Humours confusingly, from a modern scientific perspective, could also be associated with personality types or even the annual seasons. Meanwhile, the real causes of disease lurked everywhere…yet were neither comprehended nor combatted. In the words of Dr. Ann Hanson of Yale University, "There can be no question [that] infectious agents…. bacteria, viruses lurked everywhere in Rome and other ancient cities, but they went unseen and undetected." Though deeply flawed, the practice of medicine in Roman towns and cities was, however, a highly visible, generally respected, widely available and an accepted part of daily life. The Cult of Asclepius, God of Medicine, which maintained medical temples to which people could come for healing, also urged doctors to provide "poor people free treatment." Some cities and towns also furnished "public doctors" for their populations. Though technically private, the often publicly supported urban Roman medical system was often, if not always, based on doctors paid by local communities and granted tax exemptions by the Roman imperial government. In daily practice, home visits usually prevailed, and patients' family, friends, and slaves were effectively drafted into the roles of supplemental caregivers. Civilian hospitals were relatively unknown before the mid-to-late fourth century, and those epidemiological environments thus don't have to be dealt with here [4, 19, pp. 501–521; 46, 3, 33].

Most civilian physicians were general practitioners; though often higher status specialists in lithotomy, dentistry, general surgery, military combat-related surgery, parasite removal and other more narrowly defined areas existed. Midwives, physician assistants, pharmacists, herbalists, nurses, medical apprentices, and masseurs were also seen as lower-ranking medical specialists who often, if not always, operated directly under the supervision of a physician. Doctors, according to historian Dan Aparaschivei, were generally valued members of their communities distinguished by the "numerous advantages…granted to the members of their profession." Common, in-home, medical interventions included phlebotomy and other forms of blood drawing (surviving cupping vessels of a variety of sizes for bloodletting [size related to the part of the body from which blood would be drawn] testify to the prevalence of such procedures), as well as enemas and induced vomiting. Another MARV lineage R0 enhancer, in addition to these types of practices, would have been the surprising multifunctionality (that ranged beyond the modern bounds of medicine into what we would today classify as pharmaceutical, household, and cosmetic uses) of Roman medical instruments. As noted earlier, there is indeed a very strong potential overlay between such ancient practices and instances of modern medical lapses (especially in developing world settings), in relation to in-home caregiver and hospital settings, where stringent enforcement of modern sterilization/disinfection methods and protocols was lacking. Such practices all constitute strong known MARV transmission pathways; the specifics of the record-setting 2004–2005Uige, Angola outbreak are, in context, particularly pertinent here. Also applicable are Cyprian's comments, at the peak of the plague in Carthage in 252 CE, "whether they who are in health tend the sick…whether physicians do not forsake their beseeching patients, and "whether masters pity their languishing servants." Such sentiments, when viewed through the overlay of a modern medical lens, are completely unsurprising. These were all apparently very pertinent questions in Carthage in 251–252 CE. The biblical passage referenced, in this same context, by Dionysius/Eusebius…" for there is not a house where there is not one dead" is also applicable. Indeed, the possibility of such medical practices potentially turning "every house" (in the words of Orosius) into a MARV transmission setting potentially equivalent to the one which so tragically prevailed at an epicenter pediatric ward during the 2004–2005 Uige, Angola MARV outbreak is obvious [19, pp. 501–521; 32, 4, 3, 221].

A review of the common tools likely present in the standard *capsa* (medical bag) of the Roman medical profession and their uses considerably adds weight to the conclusions stated above. While Roman doctors, such as Celsus, Galen, or their third-century successors, didn't have hypodermic needles in the modern sense (that invention would only come in the nineteenth century with Scotsman Alexander Wood), they did, apparently inspired by the functioning of a snake's fangs, "use simple piston syringes to administer ointments and unctions." Such practices would have carried a high risk of mucocutaneous exposure if the instruments were *not* dedicated to a single patient (which would effectively have been a near certainty). Other medical/surgical instruments used by Roman doctors carried similar deadly risks of percutaneous or mucocutaneous MARV transmission, contaminant transfer, or other obvious, in context, epidemiological dangers. Examples include the rectal speculum, the widely used and highly varied multipurpose (medical and non-medical use) instruments such as tile cauteries, spatula probes, surgical scissors, tweezers (which were dually used for household non-medical purposes), and forceps. Similarly, physicians employed a wide range of probes, scalpels, and curettes; enema-administering clysters, a variety of sharp hooks, catheters, forceps, and chisel-like osteotomes usually (if not always) made of low-melting point bronze. Nonetheless, primitive heat-based sterilization practices, usually boiling medical instruments in water, were in use in the Roman Empire from the time of the great Methodist second-century physician Galen onwards. Yet, it is not clear how widely, consistently, carefully, or procedurally correctly, from a modern medical viewpoint, such measures were employed in urban settings. That medical instruments were rigorously sterilized to modern standards between each use seems extremely doubtful. Similarly, massive problems of, in context, infectious medical waste disposal and surface and contaminant transfer were almost certainly also not properly addressed in these prevailing home care settings. The presence of untrained slaves, friends, and family members, lack of any PPE or other modern barrier nursing procedures, the very epidemiologically dangerous nature of the common healing practices employed (phlebotomy, enemas, emetics, etc.), the risk of accidental percutaneous injury from the presence of contaminated sharp instruments, etc. are all analytically notable. From a modern epidemiological point of view, MARV's strong exposure/transmission risks in such home-care medical settings were legion. As noted above, "every house..." [3, 23, 24, 43].

This tragic R0-amplifying component in this context would indeed have been further enhanced by the highly central role of the baths in Roman medical practice and popular leisure. Baths were a ubiquitous feature of urban life. The city of Rome alone, in 400 CE had a total of 856 baths (of all types). The baths were medically promoted places of popular enjoyment, exercise, massage, health, and hygiene. Such beliefs were given a renewed endorsement in the first–second centuries by the medical encyclopedist Celsus and later by the highly influential Roman physician Galen (whose influence over Western medicine would be profound for the next millennium). Again, the overlay with potential sources of MARV transmission is striking. Celsus frequently advised the sick to visit the baths without a thought for how this might compromise the health of other bathers. People with [diarrhea]... bowel infections, for example, were urged [by physicians] to bath their exposed anuses in the public bathwater [though this particular practice would likely have been, at least somewhat, less epidemiologically dangerous with a virus than with a parasite or a bacterium] that other people bathed in." Induction of "profuse sweating" was also recommended. Such practices allegedly "ensured good health for a year or longer." Shared eye salves were also commonly prescribed in households, particularly apparently in Roman Egypt, where the hot climate caused widespread suffering from ophthalmias. Certain non-fatal, and at times not even particularly harmful, bacterial or parasitic infections that caused various forms of skin abrasions or lesions were also widely endemic within the Roman Empire's urban populations. Such abrasions or lesions would have increased the vulnerability of the empire's urban population to MARV and additionally enhanced the pathogen's R0 value in context. The best-known example is mentagra, an eruptive skin disease that was erroneously believed by the ancient Romans to be spread by kissing. In reality, it seems to have been spread by widespread Roman shaving, waxing, and epilation/depilation practices often associated with the baths. Romans treated these skin lesions through plasters and often scar-inducing cauterizations [20, p. 1; 19, pp. 52 and 455–506, 9, 3].

Such factors provide a compelling analytical backdrop for the grim urban toll and the ancient descriptions of "the carcasses of many"; the dead "left ...like refuse, unburied. The dying, "cast...out into the streets when they were half dead" in third-century Roman cities thus become very explicable, even logical when considering the potential impact of a highly lethal pathogen like MARV (with its 80% modern fatality rate and transmission profile) with the combined (often interlinked) functioning of Roman medicine and the Roman baths in an urban setting [15, p. 139 (7.22.3, 38, p. 8 (9), 10, p. 123 (16)]. Modeling the epidemiological overlay of the ancient and modern evidence paints an extremely grim picture indeed. While, despite the health benefits of the baths as touted by third-century medicine within the empire, the baths would also have been an ideal setting for the transmission of a wide variety of diseases. The potentially multiple applicable R value modifiers (infection enhancers) are particularly notable, however, in the case of a non-aerosolized, not terribly fundamentally contagious yet extremely pathogenic virus like MARV—transmitted in ways that are effectively grimly and precisely tailored to these settings and practices. There is more; baths were omnipresent in the cities of the Empire affected by the Plague of Cyprian that we have discussed– Myos Hormos, Berenice, Thebes, Coptos, Alexandria, Rome (of course), Carthage (the massive Antonine Baths), and Athens (the Hadrian Baths), etc. The baths were a quintessential part of Romanness, and even groups whose cultural and religious practices diverged from imperial third-century norms would have attended them. While groups such as Jews or Christians strongly rejected some parts of predominant Roman pagan culture, such rejection did not normally extend to the baths, "which practically everyone attended." This despite moral critiques of some of the associated behaviors—nudity, prostitution, and general sexual profligacy (in particular) [14, p. 421; 40, pp. 312–315].

It is also notable that despite a general slowdown in public works during the increasingly cash-strapped third century, emperors such as Gordian III (whose baths were "built for commoners and correspondingly equipped") and Decius (who built lavish baths for the upper-classes in Rome itself) continued to fund the construction of new baths across the empire. Even modest provincial towns would often sport multiple public bathhouses. Larger cities such as Rome or Alexandria featured a wide, sometimes huge, range of public bathing options. It is difficult to overestimate the centrality of the baths in imperial urban life." By the second century CE, any community of any substance, civil and military, had at least one set of public baths." Baths were "pleasure palaces" and places of commerce with food, drink, and sex on offer. Their "complex, luxurious, spacious" architectural layouts turned bathing into an often daily "ritual" of relaxation and enjoyment that was central for Romans [12, pp. 35 and 121; 18, 22, The Lives of the Three Gordians, (31), 40, pp. 312–315]. It is to a more detailed epidemiological modeling of this quintessentially Roman institution that we turn next.

References

1 Adams, Colin, 2007 *Land Transport in Roman Egypt: A Study of Economy and Administration in a Roman Province* (Oxford), pp. 45–50.

2 Ajelli, Marco and Merler, Stefano, "Transmission Potential and Design of Adequate Control Measures for Marburg Hemorrhagic Fever," PLoS ONE 7(12): e50948 https://doi.org/10.1371/journal.pone.0050948

3 "Ancient Roman Medicine: "Influences, practice and learning," *Medical News Today* https://www.medicalnewstoday Accessed on 12/23/2021.

4 Aparaschivei, Dan, 2012 *Healthcare and Medicine in Moesia Inferior Colection Bibliotheca Iassiensis* (Bucharest).

5 Beeching, Nicholas, Manuel, Fenech, et al., 2014 "Ebola Virus Disease," BMJ https://doi.org/10.1136/bmj.g7348

6 Beresford, James, 2013 *The Ancient Sailing Season* (Leiden, Brill).

7 Casson, Lionel, 1991 *The Ancient Mariners (*Princeton University Press).

8 CDC, 2022 "Marburg Virus Disease," https://www.cdc.gov Accessed on 7/1/2022.

9 Cilliers, Retief, 2000 "Epidemics of the Roman Empire, 27 BC-- AD 476," *South African Medical Journal,* v.90 (3)

10 Cyprian, 2013, "On Mortality," *The Complete Works of Saint Cyprian of Carthage* Translation Editing: Nicene Fathers [1885] Philip Campbell, Ed. Introduction by Ryan Grant, 2013 (Evolution Publishing,) pp. 117–130.

11 Delamater, Paul, Street, Erica, et al., 2019 "Complexity of the Basic Reproduction Number (R0)," *Emerging Infectious Diseases,* Centers for Disease Control 25, Number 1, January 2019 https://doi.org/10.3201/eid2501.171901.

12 Dvorjetski, Esti, 2017 *Leisure, Pleasure, and Healing: Spa Culture and Medicine in the Ancient Eastern Mediterranean* (Leiden: Brill).

13 European Center for Disease Prevention and Control (ECDC), 2018 "Fact Sheet about Marburg virus disease," https://www.ecdc.europa.eu Accessed on 2/4/2021

14 Eliav, Yaron, 2000 "The Roman Baths as a Jewish Institution," *Journal for the Study of Judaism* 31(4): 416-454.

15 Eusebius Pamphilus, 1890 *Ecclesiastical History* (Trans. A.C. McGiffert), Pantianos Classics.

16 Francis, Matthew, 2020 "Just How Contagious is Covid-19? https://www.popsci.com Accessed on 4/21/21.

17 Gabriel, Richard, 2002 *The Great Armies of Antiquity* (Prager), p. 9.

18 Gill, N.S. 2019 "Roman Baths and Hygiene," https://www.thoughtco.com/hygiene-in-ancient-rome-and-baths-119136 Accessed on 10/2/2020

19 Hanson, Ann, 2006 "Roman Medicine," *A Companion to the Roman Empire* (New York: Blackwell).

20 Harper, Kyle, 2017 *The Fate of Rome: Climate, Disease & the End of an Empire* (Princeton University Press).
21 Harper, Kyle, 2021 "Germs and Empire" in *Empire and Religion in the Roman World,* Flower, Harriet, Ed. Published online by Cambridge University Press. https://doi.org/10.1017/9781108932981 Accessed on 6/15/22.
22 *Historia Augusta, LacusCurtius* Trans. David Magie 1932. https://www.penelope.uchicago.edu. First accessed on 4/20/2020.
23 "The History of Sterilization Part 1," 2018 *Steris Instrument Management Services* https://www.steris-ims.co.uk Accessed on 12/21/2021.
24 Jakielski, Katherine and Notis, Michael, 2000 "The Metallurgy of Roman Medical Instruments," https://www.academia.edu/11972106/The_metallurgy_of_Roman_medical_instruments accessed on 12/23/2021.
25 Jeffs, Benjamin, 2006 "A Clinical Guide to Viral Hemorrhagic Fevers: Ebola, Marburg and Lassa," *Trop Doct* https://doi.org/10.1258/004947506775598914.
26 Killgrove, Kristina, 2016 "6 Practical Ways Romans Used Human Urine and Feces in Daily Life," https://www.mentalfloss.com Accessed on 2/2/2021.
27 Kumar, Mohi, 2013 "From Gunpowder to Teeth Whitener: The Science behind Historic Uses of Urine," *Smithsonian Magazine,* https://www.smithsonianmag.com Accessed on 2/2/2021.
28 Landau, Elizabeth, 2021"What Data Scientists Learned by Modeling the Spread of Covid-19," *Smithsonian Magazine* https://www.smithsonianmag.com/science-nature/what-data-scientists-learned-modeling-spread-covid-19-180977964/ ;
29 "Marburg Virus Survivor Holds Treatment Clues," 2020. https://www.genengnews.com/news/marburg-virus-survivor-holds-treatment-clues/ Accessed on 11/17/2020.
30 Mayo Clinic, 2017 "Ebola Virus and Marburg Virus," www.mayoclinic.org-diseases-conditions/ebola-virus/symptoms-causes/syc-2035625 accessed on 04/15/2020.
31 Mehedi, Marfique, Allison, Grosseth, et al., 2011 "Clinical Aspects of Marburg Hemorrhagic Fever," https://pubmed.ncbi.nlm.nih.gov/22046196/
32 Ndayimirije, Nestor, 2005 "Marburg Hemorrhagic Fever in Angola--Fighting Fear and a Lethal Pathogen", *New England Journal of Medicine,* https://doi.org/10.1056/NEJMp058115.
33 Nuland, Sherwin, 2007 "Bad Medicine," Review of Arikha, Noga *A History of the Humors* https://www.nytimes.com/2007/07/08/books/review/Nuland.html Accessed on 12/23/2021.

34 Olejnik, Judith, Muhlberger, et al., 2019 "Recent Advances in Marburgvirus Research," *F1000 Research*, https://f1000research.com/articles/8-704 Accessed on 9/8/2020.
35 *Orbis* Scheidel W. and Meeks E 2012: *ORBIS: The Stanford Geospatial Network Model of the Roman World*. Retrieved 20 July 2022 from http://orbis.stanford.edu originally accessed on 6/21/2021.
36 Paulus Orosius, *The Seven Books of History Against the Pagans* Trans. Roy Deferrari, 1964 Catholic University of America Press (Washington).
37 Pliny, 2003 "A Fine Crop of Poets," in Lewis, Jon, ed., *The Mammoth Book of Eyewitness Rome* (Carroll and Graff), pp. 288–299. Pliny's description is from the late 1st Century, but it would be equally applicable in the mid-3rd.
38 Pontius the Deacon, 2013 "The Life and Passion of Cyprian," The *Complete Works of Saint Cyprian of Carthage* Translation Editing: Nicene Fathers [1885] Philip Campbell, Ed. Introduction by Ryan Grant (Evolution Publishing), pp. 3–18.
39 Preston, Richard, 1995 *The Hot Zone*, (Anchor)
40 Rodgers, Nigel, 2008 *Roman Empire* (New York: Metro), pp. 284–319.
41 "Roman Ships and Navigation in Ancient Rome," 2018 http://www.vita-romae.com Accessed on 12/3/2021.
42 Schoff, Wilfred, 1917 "Navigation to the Far East under the Roman Empire" *Journal of the American Oriental Society* 37, pp. 240–249 https://www.jstor.org/stable/592924?seq=1#metadata_info_tab_contents Accessed on 10/18/2020.
43 Snelling, Nick, 2014 "The History of the Hypodermic Needle", *Well-being/Medibank* https://www.medibank.com.au Accessed on 11/11/2021.
44 Stein, R.A., 2014 "What is Ebola?" *The International Journal of Clinical Practice*, https://onlinelibrary.wiley.com/doi/full/10.1111/ijcp.12593
45 Stock, Ingo, 2014 Marburg and Ebola Hemorrhagic Fevers– Pathology, Epidemiology, and Therapy [article in German], September 2014, https://pubmed.ncb.nlm.nih.gov/1252827461 Accessed on 06/07/2020.
46 UNRV, 2022 "Roman Hospitals," https://www.unrv.com/culture/roman-hospitals.php Accessed on 7/16/2022.
47 WHO, 2014 "Ebola and Marburg Virus Disease Epidemics: Preparedness, Alert, Control and Evaluation," https://apps.who.int Accessed on 6/22/2021.
48 WHO, 2014 "Interim Infection Prevention and Control Guidance for Patients with Suspected Filovirus Hemorrhagic Fever in Health Care Settings, with focus on Ebola" https://apps.who.int/iris/handle/10665/130596 Accessed on 7/23/2021.
49 WHO, 2022 "Marburg Virus Disease," 2022 https://www.who.int Accessed on 5/06/2022.

50 Williams, Yona, 2009 "Massage Therapy in Ancient Greece and Rome", https://www.unexplainable.net Accessed on 10/20/2020
51 Zimmer, Katerina, 2020 "Why R0 is Problematic for Predicting Covid-19 Spread," *The Scientist* July 2020 https://www.the-scientist.com/features/wh-r0-problematic-for-predicting-covid-19-spread-67690 Accessed on 6/15/21.
52 Zosimus, *The New History*, Taylor Anderson, Ed. 2004 (Odin's Library Classics).

CHAPTER 8

Exploration of Modeled Urban Epidemiology Concluded and Analysis of the Contrasting Epidemiological Situation in the Imperial Countryside

It is in examining and modeling the common activities at the baths, the largest of which could accommodate thousands of Roman urbanites a day, that one sees perhaps the most concentrated potential for R multipliers for exposures to a pathogen such as Marburg (though the sum of the more spatially diffused adverse epidemiological effects of the in-home-visit-based urban Greco-Roman medical system may have surpassed the baths overall as a contributing factor to urban pestilence). Sexual activity at the baths, including prostitution, was common. As the initial 1967 German MARV outbreak proved, the pathogen can be transmitted via sexual contact (with the virus being even more persistent in the semen of survivors than in their anterior eye chambers). Though transmission through bath water itself would have been an unlikely path (particularly in the Roman *thermae*), massive communal latrines at the Baths were also potential *foci* for transmission. Communal toilet sponges *(tersoriums)* employed there (and in urban public toilets more generally as well) would have been a potential means via feces or urine. While precise details are lacking, tellingly, they were "probably reused." Certain close-quarters public lavatories attached to public baths were stifling, smelly, dark unhygienic places with low roofs and tiny windows. They were unpleasant enough that many members of the upper classes would not

M. Orsag et al., *Interdisciplinary Insights from the Plague of Cyprian*, https://doi.org/10.1007/978-3-031-26094-0_8

enter them. Physical contact through exercise, the application of exfoliating cleansers and other forms of skin care (particularly for women), as well as bath oils, massage; waxing, shaving, "hair plucking" and other depilation practices (the latter specifically also cited in the twenty-first century by the World Health Organization [WHO] as a high-level MARV transmission risk), and the use of the Roman strigil for scraping off bath oils (a behavior which could also cause sores and lesions) were all nearly ubiquitous practices at many Roman baths. Sweat, in almost all these instances, would have been the main transmission pathway. Such activities would have been specifically further epidemiologically amplified through compounding interactions with pre-existing conditions, such as widely endemic mentagra. The advice of Roman medicine for the mildly to moderately sick (such as early stage 1 MARV sufferers) to visit the Baths and engage in activities such as exercise (if possible), dry sweating (available in the *laconium* or *sudatoria* rooms at the baths) and massage would also have been, in context, highly epidemiologically problematic [9, 18, 33, 43, 47].

Massage was "massively introduced to the Roman Baths." Massage had been recommended as a cure, by no less a seminal medical source than Galen in the second century CE, for a remarkably wide range of ailments. Masseurs also often worked for, or directly with, doctors. In addition to the baths, some wealthier Romans received massages for health at home. Greco-Roman massage practices tended to be rather vigorous in terms of hand techniques with plentiful "cracks and slaps." The first century CE philosopher Seneca, who lived above a Roman bathing establishment for a while could, to his great annoyance, "hear the slaps of the pummeling… shoulders, changing its sound according [to whether] the hand is laid flat or curved." Those too poor to afford even a "cheap rubdown" oiled and then scratched themselves at *designated spots* on the walls. Seneca was also irritated by the grunts of the hard-working (and presumably also sweating) weightlifters in the bath's *palaestra*. Certain sections of the baths were typically built with specially insulated bricks [*tegulae mammatae*] that trapped heated air in these sorts of spaces. At seasonally colder times of year, patrons would have been likely to avoid the cold-water space of the *frigidarium* and be drawn strongly to these heated areas. Other areas (See Diagram 8.1 below) of the Bath complex can also likely be ruled out as risky in terms of MARV lineage transmission (apodyterium and entrance halls, libraries, and lecture halls)—in certain cases due to strong sunlight exposure (peristyles and promenade).

The settings of the hot pools of the often window-lit *caldarium* and *tepidarium*, where some sunlight was allowed to shine in to increase the heat of the water, would have been less than optimal for MARV transmission but not without some risk. Bath services (with oil application and removal) and massage areas with flying infectious sweat droplets from a feverish massaged stage 1 Marburg lineage sufferer would have essentially mirrored the initial zoonotic circumstances of the twentieth-century Kitum Cave or the Durba mine but on an industrial scale… but *sans* the *Rousettus* guano. In other words, it is fairly easy (given the techniques used) to infer not only the ease of human-to-human transmission but also the potential partial aerosolization of MARV (and thus another additional R enhancer) in such specific Roman bath massage settings. Such practices would have been analogous to but clearly, epidemiologically even worse than modern "customs involving direct touch" such as laying on of hands as part of a traditional or "cure through faith" that is "strongly implicated" by modern medical science in terms of filovirus transmission [44, 38, p. 206; 33, 9, 39]. We now turn to both macro and micro guardrail epidemiological modeling, through a systems-synthesis approach, of what we see as the probable mechanics of the urban outbreaks of the Plague of Cyprian.

Macro Urban Epidemiological Modeling: The Alleged 5,000 Deaths a Day During the Gallienic Outbreak in Rome—262 CE

Kyle Harper, in 2015, commented that it is impossible to assess the reliability of the claim from *Historia Augusta* that, in both Rome and the cities of Achaea (Greece) "five thousand men died of the same disease" in one day. This was during the massive 262 CE Plague of Cyprian outbreak. The relevant passage of the *Historia Augusta* can also be, at least plausibly, interpreted in two different ways. Though we believe, given the lack of instantaneous communication at the time, it likely means that 5,000 died in both Rome and the Greek cities simultaneously (10,000 total). One way for us to attempt to assess the veracity of this stunning death toll is to try to model, again in a rather guardrail sense, whether such an event would have been even possible in terms of MARV lineage pathology and epidemiology at the heart of the empire—the city of Rome itself. Our application of existing modeling basically assumes a Rome in

262 CE largely functioning as normal and fully populated. The decisive Roman victory at the Battle of Mediolanum three years before in 259 CE, while Third-century Crisis instability and violence persisted elsewhere in the Empire, had, after all, relieved Alamanni military pressure on the city of Rome itself [21, p. 235; 23 (The Lives of the Two Gallieni, 5.5), 1, 8] (Fig. 8.1).

Due to detailed and groundbreaking research, conducted by Drs. Marco Ajelli and Stephen Merler of the Bruno Kessler Foundation in Italy and published in 2012, we already have existing epidemiological macro-modeling of an urban uncontrolled Marburg outbreak. Ajelli and Merler's research was motivated by scientific concern in the wake of the large Uige, Angola outbreak of 2004–2005. Concerns over bioterrorism also played a prominent role as MARV "has all the key features that characterize pathogens posing serious risks for human populations if used as a biological weapon." The researchers used the largely urban Uige outbreak to establish the basic modeling parameters with an R0 of 1.59, a generation time of 9 to 9.5 days, a CFR of 88%, and a doubling time of 12.4 days. Start to peak seasonality was set as October to April (mirroring the Uige outbreak). Some gaps in the Uige data existed, and experimental primate data was used to fill them. The modeling also showed that medical early identification of potential and active cases, followed by contact tracing and

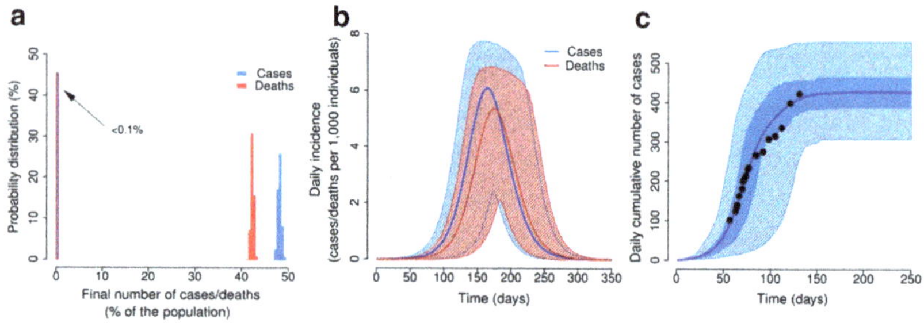

Fig. 8.1 Dynamics of Uncontrolled Epidemics and the 2005 Marburg Hemorrhagic Fever Epidemic in Angola. The blue line indicates modeled daily incidence of new cases (dashed area represents mean and 95% confidence interval [CI], respectively); the red line indicates deaths (dashed area represents mean and 95% CI respectively) of simulated uncontrolled epidemics, starting with one infected individual at time 0 in a population of 100,000 individuals. Figure courtesy of Dr. Marco Ajelli and PLoS One (2012)

case isolation or, alternatively, a strategy of "intensive social mobilization," as indeed had happened in Angola, were generally effective strategies in containing large MARV outbreaks [1].

In the third century, obviously little of that would have been possible. The outbreak would have been relatively uncontrolled. The modeling, with a population size of 100,000, was based largely on pathogen characteristics in an urban environment. It did not, however, account for population density specifically. Peak infections would occur on day 167 of the outbreak, with deaths lagging behind (day 177). A noticeable level of abatement would naturally begin at around 220 days. The full outbreak, upper range epidemiological parameters for final attack rates would be approximately 48% of the population. Consistent with the pathology of a CFR of 88%, the virus could kill up to 42% of the modeled population. Though before that grim scenario, abating social distancing would likely be "spontaneously driven by the reaction of the population to an ongoing [terrifying and deadly] epidemic" and "would be performed with non-negligible impact." Much of what we read seemed eerily familiar to us when we analyzed Ajelli and Merler's modeling rather late in our research process. Overall seasonality and duration fits very well with what we have deduced for the Plague of Cyprian's outbreaks in Carthage and Alexandria. Around six months from outbreak to peak deaths, the matching specific and broader seasonality timing (as in Symeon's) also fit remarkably well, as does the occurrence of a, to some degree epidemiologically concluding, partial or full *ville morte* event. The overall analytical impacts of these congruences are, of course, partially diminished by a key potential difference in terms of human-to-human transmission engendering the referenced third-century outbreaks and zoonotic transmission, albeit not specifically identified, assumed to be initially responsible in the case of the modern Uige outbreak. The modeling also describes the inherent rather unpredictable stochasticity of MARV outbreaks, which can grow to reach truly horrific levels or just as easily die out during their fragile epidemiological beginnings in particular [1].

A rather simple mathematical extrapolation of ours (using Hanson and Ortman's 2017 estimate of Rome's population size at 920,000) from the Ajelli–Merler modeling also, at least largely, answers the question of whether a MARV lineage outbreak could have been theoretically responsible for the reported ancient death toll of 5,000 a day in Rome. Less than two weeks before peak deaths, daily fatalities would have been between 2,000 and 2,500. Our calculations put the single daily peak death toll for a Rome outbreak at 4,876. Accounting for Rome's extremely high population density of 513/ha in the modeling, ancient hyperbole or a combination of urban Roman and Greek death tolls (the wording as noted is ambiguous) could each easily more than account for the remaining small gap. Before (deaths would have been an epidemiologically lagging indicator), during or immediately after the advent of such a horrifying death toll, panicked flight from the city (well-documented in the Carthaginian and Alexandrian cases) would have likely begun with an ultimately strong epidemiologically mitigating effect on the urban outbreak though also producing epidemiological dissemination to surrounding areas [20, p. 319; 1].

Micro Urban Outbreak Modeling: The Spatial Epidemiology of the Baths of Caracalla

We also wanted to interrogate and demonstrate some of the potential spatial epidemiological exposure parameters for the largest and most lavish

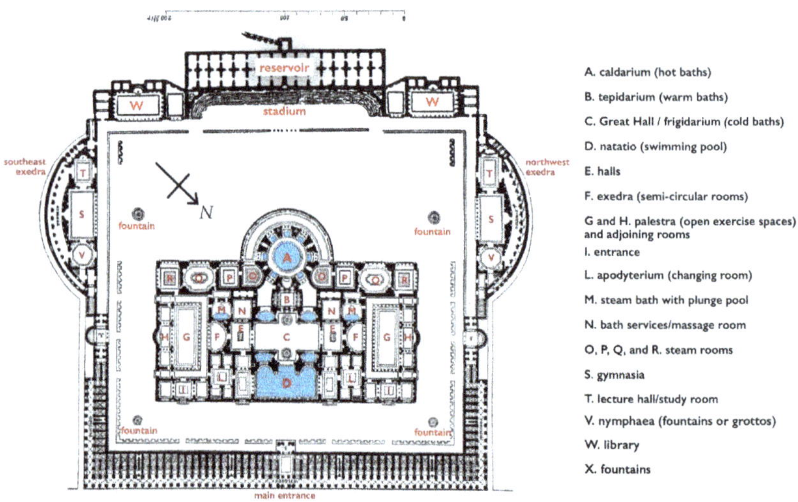

Diagram 8.1 The Baths of Caracalla. Created by B. Fletcher [public domain]... and as rendered by Jessica Mingoia, "Baths of Caracalla, Smart History, *The Center for Public Art History*, 2022 [31]

public bath in Rome during the mid-third century, the Baths of Caracalla. To do so, we modeled five Romans of differing genders, ages, and social statuses (this modeling was not meant to be a demographic reflection of the city's population or a precise reflection of Roman sociocultural practices regarding mixed [by gender] bathing) but rather a theoretical epidemiological selective modeled sample designed to examine different patterns and levels of MARV lineage exposure risk at a very early stage of an outbreak in the early 250s or 260s CE in the city of Rome. The overall results are summarized in Table 8.1 (based partially on Diagram 8.1). All modeled infected entered the Baths in the afternoon during the busiest time of day. Four were at different intervals of the first two days of stage one MARV pathology with one not being very sick at all yet (headache only), through two also suffering from considerable headache, malaise, and myalgia; to one with more severe debilitation, a splitting headache and beginning to suffer from bouts of severe diarrhea. She needed assistance from a friend/fellow household slave to enter. A fifth person was in the beginning interval of stage 2 pathology, and was suffering from even more severe debilitation (he had to be helped/carried into the baths by two slaves) and additionally from intermittent but violent nausea and vomiting. We applied the exposure criteria from Dr. Matthias Borchert's epidemiological study of the 1998–2000 Durba-Watsa Marburg outbreak with the addition/intensification of a (somewhat lesser) originally implied exposure risk from sweat. We believe this alteration to be medically/epidemiologically justified given the particularities of some of the spatial epidemiological environments present at the baths and specific vulnerabilities endemic to the Roman population—e.g., widely endemic mentagra. Borchert's study indicated that mild or asymptomatic MARV infection occurred during the 1998–2000 outbreak but was very rare. We thus didn't attempt to account for that factor in our modeling. Borchert's original MARV exposure criteria were as follows (below Table 8.1):

Depending on symptoms associated with increased virus shedding and the intensity of contact, level of exposure was categorized. Low-level exposure contact included any direct contact with a living case patient without diarrhea, vomiting or bleeding; or touching clothes or sheets not soiled with stool, vomitus or blood. Medium-level contact was defined as touching a living case patient with diarrhea, vomiting or bleeding; touching clothes or sheets with stool, vomitus or blood: or touching remains. High-level contact included carrying or embracing a living

Table 8.1 Modeled MARV Lineage Spatial Epidemiology in the Baths of Caracalla (Rome) by Dr. Amanda McKinney, Dr. Mark Orsag and Melissa Clouse

See Diagram 8.1		*Exposure level*			
		High	*Medium*	*Low*	*Total*
Location in the Baths of Caracalla	I	0	0	0	0
	L	0	0	0	0
	T	0	0	0	0
	W	0	0	0	0
	G	0	0	0	0
	H	0	0	0	0
	S	0	0	0	0
	N	**4**	**8**	**1**	**13**
	OPQR	**3**	**2**	**6**	**11**
	M	0	**1**	**1**	**2**
	A	**3**	**1**	0	**4**
	B	0	0	0	0
	C	0	0	0	0
	D	0	0	0	0
	E	0	0	0	0
	F	0	0	0	0
	V	**1**	0	0	**1**
	X	0	0	0	0
	LF (two types)	**4**	**3**	0	**7**
	Total	15	15	8	38

patient who had diarrhea, vomiting or bleeding; or carrying, embracing or cleaning remains [8].

Many areas (see both Diagram 8.1 and Table 8.1 above) of the Baths of Caracalla were effectively zones of no to little epidemiological risk. We also assumed no risk levels for the archaeologically missing (and therefore mysterious) second floor of the baths. Most of our modeled Romans were already too debilitated to engage in vigorous exercise. The least afflicted one, however, did manage some light exercise (without measurable modeled exposure to others) and also engaged the services of a prostitute—a high exposure. One type of very high-risk spatial area (N) included both the bath services and massage rooms. This was due to the extended close contact involved in massage (something, however, that not all of our modeled Romans could afford) and/or application and removal of bath oils. The differing steam rooms (O, P, Q, R), and to a lesser degree the *caldarium* (A), which we modeled as collectively very hot and tightly packed despite the latter's 130-foot dome) also were zones of relatively high exposure risk to other bathers and the staff from our five modeled infectious visitors to the baths. The two different accompanying types of toilets (not shown in the diagram) collectively constituted another high-risk area. The answer as to how many new cases would likely have arisen, in an average of 9 to 9.5 days later (Tg) from these collective modeled exposures, however, is one that we ultimately chose not to try to answer to a high degree of specificity. Borchert and his colleagues modeled a collective secondary attack rate of 21% (interestingly the tertiary attack rate was barely lower at 19%), but his study didn't fully distinguish between varying degrees of exposure in this sense. The actual exposures and resulting cases uncovered by Borchert, through serological testing and contact tracing, following the 1998–2000 outbreak in the DRC, also tended to be somewhat more strongly skewed toward high exposures than our modeled ones. Countervailingly, Borchert's study charted a period of the Durba-Watsa outbreak post-intervention; the R0 value had already been reduced to 0.93. For the early stage of the devastating Roman outbreaks of the early 250s or 260s CE, the reproduction number would have been well above 1.0. Additionally, our modeling was *not* demographically representative. Thus, in light of that confusing mix, we only modeled exposures as opposed to actual projected resulting cases—though the overall projected epidemiological driver threat from merely a single day of modeled bath operations was obvious to us [8].

The overall scale of how the virus' R0 would have been affected by the above-demonstrated patterns of its probable interactions within Roman urban society would likely have produced the very "slow burn to steep build" epidemiological third-century urban profile that the Plague of Cyprian demonstrated as infections, through cascading exposures, would have slowly built until such multiplication (through the related medical and bath "super exposers") became truly epidemiologically significant.

A third major modeled socio-cultural urban R0-amplifier would have been third-century funerary customs within the Roman Empire. Funeral rites were particularly important in that predominant pagan religious beliefs taught that those buried without them were condemned to roam the earth as ghosts—haunting the living. Cemeteries were also located outside the boundaries of towns and cities. Yet, customs and practices varied greatly by social class and era. During the mature Roman Republic and early empire quicker, somewhat simpler, but more costly to arrange, cremation prevailed. By the third century, however, the socio-cultural pendulum had swung back to longer and more elaborate burial practices. Elaborate funerals for political and cultural elites contrasted, however, with the quick dumping, and leaving for decomposition, of the bodies of slaves, criminals, and the indigent outside (legal) or inside (illegal but not uncommon) city or town limits. Infant and child mortality was also very high and similar practices prevailed in those cases. Burial in the catacombs under the city of Rome itself was also an option. The tombs and mausoleums containing the sarcophagi of the wealthy, by contrast, lined the roads approaching imperial towns. Yet, sporadic state support for the funerals of less well-off urban citizens and burial societies (organized commercially and/or by guild/profession) allowed many less wealthy citizens to at least attempt to imitate the more elaborate funerary practices of wealthy elites. Many burial societies were open to anyone, even slaves, who could pay. A class of professional undertakers existed. Costs for funerals had increased by the third century both due to more elaborate ceremonies and chronic inflation. The wealthy, and those to some degree imitating them, often lay in state in their homes for up to a week. Adornment, painting, and cleansing rituals occurred in which female relatives and the undertakers handled the corpse. Embalming was relatively uncommon outside Roman Egypt but far from unknown. During plagues and pandemics like the Plague of Cyprian, the complex urban system of funeral, burial, and body disposal could completely collapse with carcasses of citizens of all classes simply left in the street to rot—something, as noted in Part I, described by numerous ancient witnesses in the case of the POC [6, 7, 42, pp. 43–45; 34].

Rural elites imitated the elaborate funerary practices of their urban upper-class counterparts in the cities and ended up, similarly, in rural family mausoleums. Their similarly epidemiologically risky rites and funerals would have been, however, much more spatially and temporally separated in an epidemiological sense. We know relatively few details about the burial practices of the rural poor (Christian or pagan), but their funerals seem to have been comparatively quick and simple. Bodies, were buried, within a day, with a few grave goods, in simple pit graves dug in ground deemed not worth planting—located near the fields that they had worked in life. Estate owners permitted but generally did not pay for such burials. Land for such burials in low-population-density rural areas was thus not in short supply [42, pp. 43–45; 34].

Christian rites and burial practices during the third century often differed only minimally from those of the hegemonic pagan culture. Christians imitated Greco-Roman pagan funeral rites and customs in many respects. Christian sarcophagi, for example, displayed artwork that only covertly differed from those of pagans. This can be explained both by cultural hegemony and an, often well-founded, fear of potential persecution. A more overt and independent socio-cultural identity for Christianity would, for obvious reasons, emerge in the fourth century. Regional funeral customs within the vast third-century Roman Empire, by contrast, displayed considerable variation and a pattern of cultural hybridization between local and prevailing Greco-Roman cultures. As we have already discussed, Roman Egypt serves as a particularly useful example. Egyptian burials were usually far from prompt. Varying degrees of careful preparation (including preservation, washing, painting, wrapping in linen, and adornment of the body) were the norm. The body could be kept above ground for days or even weeks. Embalming and mummification, sometimes (in an example of the cultural blending of traditions) with the addition of a Roman-style death mask to the mummy were also still commonly practiced. Massively intrusive mummification, with a wide range of cost options available, was indeed open to all who could afford it. Yet, inability to pay for even the cheapest mummification option was still a major problem for the poor; they made do with simpler desert burials that still involved washing and wrapping the body in cloth. Even during the funeral, itself, the body was often touched by the presiding priest and/or family members. This range of Egyptian burial practices, like the Greco-Roman ones, raise an ascending hierarchy of modern epidemiological "red flags" in relation to the POC. The collective modeled epidemiological and pathological outcomes of such practices also fit perfectly, in an inferential sense, with the kind of evidence detailed in Part I relating to the Funerary Complex of Harwa. The later Byzantine accounts of George the Monk and George Kedrenos also noted disruption of burial customs during the Plague of Cyprian—another confirming sign that supplements the accounts of Cyprian and Dionysius in terms of the epidemiological dangers (and Romans' unknowing but effectively mitigating reactions in response) lurking, in context, within these customs. Additional archaeological confirmation of such disruption may also eventually be provided by the aforementioned Catacombs of Marcellinus and Peter [41, 32, 21, pp. 226–233].

While the holistic verdict of our epidemiological "guardrail" modeling for the MARV lineage pathogen estimated the rural threat embodied by such end-of-life practices as fairly moderate, the overall R0 estimate for Roman third-century urban areas was once again, contrastingly, highly amplified and exceedingly grim. Given complete Roman ignorance of the processes involved in viral transmission and the relative paucity of source material for the third century, it is surprising that any potential corroborating evidence of our urban analytical deductions exists. Yet, specific ancient source evidence from four differing points of origin seem to fit very well with our modeling's designation of the baths, the Roman medical system, and extended funerary rites as the major urban "super-exposer" epidemiological *foci* of the Plague of Cyprian's MARV evolutionary lineage pathogen—with the specific practice of massage directly implicated in one instance [36, pp. 13–47; 40].

Plotinus, Cyprian, Dionysius/Eusebius and the Re of Hanina bar Hama

As noted in Dr. Kyle Harper's corpus of work on the Plague of Cyprian, the late third-century Neoplatonist philosopher Porphyry recorded that an earlier Neoplatonist sage named Plotinus had survived the plague though multiple masseurs that had formerly worked on him (at home) later became infected. Plotinus apparently gave up massage and avoided the baths during the outbreak; thus, he is another good example of the "rough and ready" Roman awareness of disease communicability among educated Romans. Plotinus survived. The masseurs involved, however, weren't so lucky. They all died. It is probable that they had also plied their trade in the baths. While not conclusive, MARV transmission through sweat from stage 1 (or even stage 2) sufferers seeking relief, on the advice of Roman medicine, at the baths and through massages fits this instance very well. Plotinus' example is also a reminder that urban depopulation, military and civil violence and economic and security disruption due to a variety of causes (including the plague itself) during the Third-Century Crisis may have worked, short term, to spread the Plague of Cyprian locally, but also likely were ultimately factors in containing it. This epidemiological abatement is due to the reality that the interrelated urban "super-exposer" institutions of funerary customs, the medical system, and the baths would have ceased to function fully. Aparaschivei notes, for example, that the operation of provincial military hospitals and urban medical temples in a number of regions of the empire was, at least temporarily, partially or completely disrupted as the Third-Century Crisis worsened [21, p. 237; 29, 3, pp. 104–106]. If conditions improved, however, while the virus persisted at some level, renewed functioning of institutions such as the medical system and the baths would potentially again lead to a repeated "urban build" pattern. Cycles of behavior regarding funerary practices would likely have functioned similarly in an epidemiological context. This would partially explain (along with the [particularly maritime and riverine] interconnectedness of the empire) why certain cities, such as Rome, were apparently hit twice and why, after the initial ever-expanding wave of the plague that ravaged the empire through the early 250s CE, large outbreaks of the Plague of Cyprian seem more sporadic during the late 250s and early 260s CE. Though other unrelated potential causes, such as more limited source material generation, disruption due to increasing barbarian invasions, political fragmentation, economic breakdown, and civil wars also offer logical alternative inferential explanations.

Rabbinic literature and specifically Hanina bar Hama, who was discussed extensively in Part I, provides another evidentiary confirmation in this regard. In this case, the source material is more complex and less straightforward, yet its overall impact is uniquely striking, in terms of both its breath and precise fit when considered in light of our overall conclusions. As noted, despite religious/moral strictures and concerns involving mixed bathing and other forms (such as prostitution) of licentious behavior, most Jews used the Roman baths. The. The baths were, after all, "essential to people's way of life" in the Roman Empire. Their health benefits were also widely accepted by Greco-Roman medicine, which was also widely utilized by the Empire's Jewish residents. Hanina bar Hama, however, was something (if not a total exception) to both of these rules. As noted in Part I, the third-century Talmudic Sage was also a physician and oil merchant whose non-standardized beliefs about medical issues were clearly well known among the Jewish community in his hometown of Sepphoris. The rabbi's eschewal of established Greco-Roman medical practices would not necessarily have dramatically reduced his (or that of his family and associates) exposure to many pathogens, particularly airborne ones. Yet, if we again examine Table 7.2 (related to MARV exposure) in relation to his beliefs, whole rows of potential transmission pathways begin to disappear [14, 30, pp. 192–193; 46 (14b), 5 (107b, 24b)].

At least in large part, Hanina bar Hama also seems to have avoided the traditional Roman baths. He nonetheless stressed the health benefits of hot baths (again something that his mother had done for him as a child). Rabbi Hanina, however, did bathe at the therapeutic sulfurous thermo-mineral hot springs (baths) at a renowned place called Hammat-Gader (which is still a major tourist site in modern Israel today). Hammat-Gader was famed for its medicinal properties throughout the empire, and the Romans constructed an "impressive and monumental" bath complex at the site of one of the five springs at Hammat-Gader. While the buildings and an overall Roman bathhouse atmosphere prevailed (with hospice areas, food stalls, shops, and even a theater), Hammat-Gader was (at its core) a natural, as opposed to man-made, facility. It was based less on the traditional Roman bathing rituals (with their sequence of cold and hot baths) and thus lacked some of the "functional elements" and practices common to most Roman bath complexes. Hammat-Gader's bathing experience focused more on the "actual perceived properties of the water" in the hot mineral baths. Thus, a number of the more problematic (in an epidemiological sense re MARV) Roman bath-based activities, such as massage, exercise in the *palaestra*, use of the "dry sweat" rooms, listed in Table 7.2 seem not to have been practiced there. The facility also seems to have lacked both a *caldarium* and any sign of a *hypocaust*. The clientele was, however, numerous, multiethnic and multicultural, and also probably often included legionnaires from the 10th *Fretensis* (which was based nearby). A licentious atmosphere prevailed and many other Roman bathing practices and traditions, including mixed bathing (which flourished despite the attempts of multiple emperors to suppress the custom at the site) were still followed in this more natural setting. Sages such as Hanina bar Hama found the site's "decadence objectionable." Yet, the perceived health benefits of that "certain place" (as Hanina bar Hama, with deliberation, obliquely referred to Hammat-Gader) outweighed its association with, among other things, prostitution. As the historian Dr. Esti Dvorjetski has noted... "Their [the Hammat-Gader Thermo-Mineral Springs] use by the Jewish inhabitants created a fascinating conflict between the popular need, especially for healing, and the religious restrictions, which were discussed and decreed by the Talmudic sages." It is within this cultural context that Dvorjetski provides details of Hanina bar Hama's activities at Hammat-Gader [12, pp. 280–307, 28, 121–152 and 367–370].

The rabbi and those close to him seem to have engaged in bathing, some commercial activities, and lengthy face-to-face meetings to discuss questions of *halakhic* law and rulings with a number of other visiting Talmudic scholars from the regions around Hammat-Gader. They also sampled the products of the site's cookshops in a way that also reflected traditional practices of the Roman baths that Seneca, among others, had noted; all fairly low-risk activities in relation to a non-aerosolized viral pathogen such as MARV (though *notably not low risk at all* for many other types of aerosolized [such as influenza, measles, and smallpox] and various other non-aerosolized diseases and infections). The rabbi additionally practiced and preached religious strictures that curtailed and shaped what was permissible at Hammat-Gader. Contacts, outside a tight known Jewish circle, and activities were to be limited due to both moral and practical reasons. "Take note that there are people there who are invalid," Hanina bar Hama decreed. He further observed what could happen if such strictures weren't observed (and apparently many Jews didn't observe them), "...many children of non-Jews were born to Jewish women." While at Hammat-Gader, sages such as Hanina bar Hama also seem to have covered (for reasons that are not entirely clear) any existing skin lesions or injuries with a special type of bandage called an *ispelanit*. "This bandage... was covered with viscous material which stuck fast to the body and did not melt as a result of... heat." Rabbi Hanina and his circle also seem to have kept mostly to themselves and had their own separate bathing practices. Traditional Roman bath methods of oil application and/or removal (such as the strigil) were eschewed in favor of removal via smooth pebbles that Hanina bar Hama and his close associates carried with them and used at Hammat-Gader. Given the sage's "unconventional" medical views and assuming that Hanina bar Hama and those who followed his rules in general practiced what they preached at Hammat-Gader (and there seems to be no evidence to the contrary), the whole of our Table 7.2 would have been essentially inapplicable to the sage and his close followers. No cascading repeated MARV lineage exposures for or from them [12, pp. 296–307; 38, p. 206].

Reduced risk of transmission for Rabbi Hanina and his followers can also be found in the funerary practices associated with them as opposed to more generally in the third-century Roman Empire. Traditional Jewish customs, originating in hot climates, stressed "prompt burial" with the corpse interred on the day of death or, at most, a day later. Ancient Jews "did not embalm the dead." There was indeed a fairly strong religious tradition against keeping a corpse above ground overnight. Some funerary practices would, however, have entailed considerable risk of exposure—such as washing the body with perfumes and ointments and wrapping it in cloth. Again, however, a more detailed examination reveals factors that would have mitigated personal epidemiological exposure from such practices for Hanina bar Hama. The third-century *Tractate Semahot* of the Babylonian Talmud, which details approved funeral rituals, makes clear that women were allowed to prepare both male and female corpses for burial while men were only allowed to prepare the corpses of men. The third century additionally witnessed the cultural origins of the *Chevrah*—mutual specialized burial societies—organizations of men and women (referred to as *Chevra Kadisha* in modern times) charged with seeing that dead Jews would be buried according to Jewish rites and traditions. Other members of the Jewish community would not be as involved in such preparation and carry on with life and work. Any corpse with visible blood would also be handled less but rather remain in any blood-soaked garments, over which white sheets would be placed before transport to the burial site. Rituals generally were supposed to be simple and quick—followed by transport to family burial sites located in soft rock caves. The "second burial" (partial reburial to an ossuary and/or common inner sanctum) carried out at the cave sites a full year later would not have risked exposure as the RNA virus would have been rendered inert long before [11, 17, 28].

Overall or regional Greco-Roman (also Greco-Egyptian) cultural hegemony in the third-century Roman Empire initially seems to complicate this narrower analytical portrait in this regard, but closer examination actually reveals another remarkably precise fit for a key piece of the puzzle. Many Jews began to follow the more elaborate, epidemiologically risky in context, prolonged, and costly preparatory practices of surrounding majority cultures that at times included embalming and use of expensive and elaborate coffins and even sarcophagi. Many Jewish funerals became prolonged and "costly affairs with increasingly extravagant spending on adornment of the dead." In fact, in the views of some scholars, truly "distinctively" Jewish/Israelite funeral/burial patterns for the period are difficult to document. Yet, once again, Hanina bar Hama's religious beliefs would have fortuitously served as exposure mitigators, both for himself and for faithful followers. Rabbinical followers of Hanina bar Hama's mentor Rabbi Judah made very limited concessions (only simple coffin use was permitted) to such trends and generally strongly opposed them. Such a lengthy, elaborate, costly, majority-culturally influenced, and expanded Jewish funeral rite was seen by these rabbis as a sacrilegious, alien, and "ostentatious foreign influence and [they] rejected it [11, 17]."

As noted in Part I, Hanina bar Hama, still a respected (if at times controversial) religious elder, seems to have survived the Plague of Cyprian in a way that characteristically attracted attention in his hometown of Sepphoris. As Stuart Miller, whose article "Hanina bar Hama at Sepphoris" analyzed rabbinical source material noted, the ancient sources recorded that...." There was a pestilence in Sepphoris that did not spread to the street on which R. Hanina lived. What does this elder lose? For he sits in health together with his neighborhood while the city perishes in illness." Some in Sepphoris even seemed to blame the famous religious sage and physician for not curing the plague. Again, if taken literally (and one does have to approach such a source with a degree of caution) this seems to be a rather unlikely eventuality with an airborne or other more easily transmissible pathogen. If one makes the quite limited and supportable further logical inferences that the disease outbreak discussed was, in fact, the Plague of Cyprian (and indeed it easily fits best in terms of chronology as there seems to have been no recorded major epidemic disease events in the decades between the final waves of the Antonine Plague [when Hanina bar Hama would hardly have been well-known "elder"] and the outbreak of the Plague of Cyprian), and that Hanina's strictures and overall socio-behavioral/religious rulings would probably have been most influential with his immediate neighbors, Hanina bar Hama himself serves, for a remarkable second time and in multiple applicable contexts, as a highly notable analytical puzzle piece in terms of our overall proposed solution to this third-century pandemic [30, pp. 192–193; 16, p. 139 (7.22.6); 22, p. 136].

But with its terrifying symptoms and a modern average death rate of 80%, the fear and shock that the novel pathogen would have generated in third-century Roman cities would almost certainly have been exactly embodied by Dionysius' surviving (through Eusebius) description, in terms of the disease assailing the broader non-Christian community, ..."More dreadful than any dread, more intolerable than any calamity... the only thing that prevails over all hope [16, p. 139 (7.22.6)]." There is a modern overlay with our overall modeling of the Plague of Cyprian pandemic. The West African Ebola Zaire outbreak of 2013–2016 killed 11,325 people across the nations of Guinea, Sierra Leone, and Liberia. Modern medical intervention finally brought the outbreak under control after 28 grueling months. As dozens of cases spread also across "seven other nations on three continents," the event technically/definitionally thus qualified as a modern filovirus "pandemic [39]."

The Plague of Cyprian in the Rural Empire

At this point, in order to wrap up some unfinished business from Part I's List 2.1 Point #22, we must shift our focus from third-century Roman cities to the empire's villages and rural areas. The R0 environment, including in terms of population densities, would have been dramatically different. The Greco-Roman professional physician caste of town and garrison would also have been largely absent as would larger public baths. Much of the medicine practiced in these areas would have been of the enduring folk as opposed to the prevailing Greco-Roman variety. Detailed accounts of the Plague of Cyprian in these rural settings, such as those of Dionysius/Eusebius, Cyprian, and Pontius for the urban areas, are nearly completely lacking. While as noted in our Table 7.1, ancient sources such as Orosius and Zosimus do, in a very brief and general sense, indicate that the Plague of Cyprian did hit the rural areas of the Empire hard, and Kyle Harper seems to (at least broadly) accept this in his pioneering analysis, our own verdict is far more doubtful on whether rural areas were anywhere near as heavily impacted as urban settings by the Plague of Cyprian. Orosius and Zosimus may have, we surmise based on the overall verdict of our modeling, either been engaging in exaggerated or ideologically/religiously motivated speculation (Orosius) and/or been disproportionately influenced by *ville morte*-influenced epidemiological dissemination from hard-hit cities into the rural areas immediately surrounding them. Our overall interpretation here is different from Kyle Harper's and far more in line with the assessment (in terms of more epidemiologically limited impact and dissemination) of Sabine Huebner. We believe the impact of the Plague of Cyprian in rural areas was likely to have been somewhat limited (though obviously not nonexistent). Given the silences and vagaries in the ancient sources, generalized hyperbole, to us, seems highly likely here. Our proposed urban Plague of Cyprian modeling, particularly in terms of transmission pathways, also fits far less well with less densely populated rural areas without the urban presence of the linked "super-exposer" massive Ro enhancers of the baths, funeral customs and the Greco-Roman medical system. Our modeling takes the plague from urbanized areas of Aksum through largely maritime means, to the Roman Empire and across the Mediterranean and mostly downstream in terms of riverine travel. The aforementioned "constraints" and "frictions" of ancient travel—time, expense and distance applicable to ancient pandemics, noted by Harper himself in his analysis of the Antonine Plague, also seem definitely applicable. As Harper noted, "The experience of the pestilence in a village, or in an army barrack, or in a metropolis, would have differed." [22, p. 140].

On the other hand, the fact that the Plague of Cyprian's pathogen spread to rural areas to some extent seems nearly certain. Inferentially, de-urbanizing plague-driven exoduses—the "rough and ready" perceptions at work again—from stricken cities, such as the one noted by Dionysius/Eusebius for Alexandria, would have at least spread the plague to the surrounding countryside. Later in the Third-Century Crisis (in the early 260s), barbarian invasions and sacking of Roman cities may well have had similar R-multiplier epidemiological effects, as these were effectively also *ville morte* events. The fact that the two most prominent ancient sources in this regard, the Christian Orosius ("no Roman province, no city, no house"') and the pagan Zosimus ("cities and villages"), are from sharply differing ideological/religious "camps" is (indirectly) a point in favor of their, albeit totally non-specific and extremely general and vague, claims of nearly universal pestilence that afflicted both cities and villages—apparently equally. On the whole though, the rural areas likely demonstrated, at worst, a far more limited "slow burn" pattern and scale more characteristic of modern MARV or EBOV outbreaks in rural areas of Africa (sans the modern medical intervention) as opposed to the terrifying "build" graphically reported in ancient cities such as Alexandria and Carthage. Lower ancient rural population densities and a complete lack of modern means of mobility (technology as simple in the modern age as a bicycle) would very likely have caused many outbreaks outside of more urbanized third-century epidemiological environments to flame out quickly due to a lack of sufficient contacts to keep R0 values above 1.0. High levels of commerce and movement in terms of waterborne travel would also have been far less of an epidemiological contributor for rural environments, on average, than for urban areas (disproportionately situated on coasts or on rivers). In third-century outbreaks, unlike in the two-yearlong Durba-Watsa outbreak in the DRC, there would, however, have been no effective medical intervention. Counterbalancingly, however, there also would have been no chance of "multiple introductions of the virus into the human population through wildlife" as occurred in that case. Contrastingly, the largely urban Uige outbreak occurred after "a single introduction of the virus in the community was responsible for the epidemic." Human-to-human transmission took over, in an epidemiological sense, from there. But to temporarily play devil's advocate to our own conclusions... If the Plague of Cyprian did heavily impact "cities and villages," as indeed directly claimed by ancient sources as opposite in outlook as Zosimus or Orosius, how could that have worked, in terms of the required R0 enhancement, with our solution's implication of the MARV evolutionary lineage [48, p. 8 (1.15–16); 35, pp. 315–316 (7.20–22); 2]?

While our analysis and modeling here are somewhat more speculative, there are some interesting potential epidemiological dissemination paths that might possibly account for more widespread transmission of MARV in third-century Roman village/rural settings or at least explain how more limited outbreaks might have functioned in terms of epidemiology. Only a few, in context, epidemiologically risky practices such as the shift toward more elaborate funeral rites would have been shared between third-century Roman urban and rural settings (and these likely applying not universally but only along class lines). Even in that instance, cost and logistics would have been a mitigating factor along with lower population density for rural areas. The three main urban super-exposer institutions would thus have been largely (if not fully) absented in rural settings. One possible MARV dissemination pathway more specific to rural settings would have been certain well-documented practices of Roman folk medicine. While the Greco-Roman medical corpus had a logical and systematic, if erroneous, coherence due to its overarching four humours theory, Roman folk medicine practices, not unknown in urban areas but far more widespread in rural areas, seem to have been a largely inchoate mass of strange ("...eat bites from the liver of a dead gladiator" was one such recommendation) superstitions. Along those same lines, we noted the Roman folk medicine superstitions regarding bat organs and tissue in Part II of this series. Though some of these sorts of practices were apparently fairly widespread, regional variations would have also probably placed a certain upper limit in terms of how much pan-imperial enhancement of non-aerosolized MARV's fairly low base "biologic R0" they could have actually provided—though such impacts could still have been locally (even regionally) significant. Transmission via medical use of urine is one relevant path. Folk medicine advocates recommended bathing in "cabbage urine" (the urine of someone who ate cabbage frequently) both as a prophylactic to ensure general health and as a "cure-all" for a wide variety of, including particularly gynecological, ailments [19, pp. 498–508]. Table 7.2's row 11 is applicable here; though only if treatment aging processes did not kill the virus; something which is unknowable (but seems inferentially very far from improbable).

Human feces and urine were also utilized as part of the Roman Empire's "waste not want not" agricultural economy in rural settings for fertilizer and veterinary purposes. Human urine was seen, in particular (as noted by the first century CE agricultural commentator Columella), as a key ingredient in pomegranate cultivation. Such traditional methods would almost certainly still have been in use in the third century. Usage in veterinary medicine seems to have been an outgrowth of Roman folk medicine practices (discussed above). The range of animals treated with urine was wide and included mammals, birds, and even bees. As with our ancient–modern evidentiary urban overlay involving these human waste products, however, the revelatory epidemiological picture is cloudy, holistically limited, and somewhat imprecise. Collection, treatment, storage, and utilization processes regarding such human waste products in Roman agriculture aren't precisely known. Some of these may likely have rendered the pathogen inert when such human waste products were utilized for veterinary or fertilization purposes. Some limited, both individually and in terms of epidemiological scope, risk to early stage collectors/handlers seems a defensible inference, but there seems little upon which to base speculation of the sort of "super-exposure" R0 enhancement akin to that of the functioning of the baths, the medical system and complex funerary rites in high-population density imperial urban settings [26].

A third highly intriguing, yet even more speculative, pathway for MARV transmission in imperial third-century rural settings would have been "reverse zoonosis" followed by zoonotic re-infection via a truly ubiquitous fixture of the Roman countryside and village life—the pig. The phenomenon of pigs as strongly suspected amplifier hosts in twenty-first century Sub-Saharan Africa MARV outbreaks was discussed earlier in Part II. But without Egyptian rousette bat populations for the pigs to potentially interact with across most of the expanse of the third century Roman Empire, another transmission pathway must be postulated: Reverse zoonosis is indeed an increasing phenomenon in the twenty-first century as a result of intensifying human-animal contacts. Animal husbandry of swine in the third-century Roman Empire was similarly widespread and characterized by "intense cultivation." Widely available and often relatively cheap, pork was the one meat that Romans of all social classes seem to have eaten fairly regularly. Its widespread production seems to have weathered the Third-Century Crisis.

The anonymous fourth-century source the *Epitome de Caesaribus* mentions Emperor Aurelian (*Orbis Restitutor*) having provided (toward the end of the Third-Century Crisis) "for the [general] populace... a ration of pork." Consumed by the rich and poor, it was often considered the "choicest of domestic meats." Romans had at least fifty different ways of cooking pork. Perusal of Apicius' late fourth-century cookbook *De re coquinaria* bears this out; it produces recipe after recipe and demonstrates the prominence of pigs in the Roman diet—for instance, the eight different types of sausage meat listed in Book II. These include multiple recipes for bacon and "diced, cooked pork shoulder." Pork could be fermented in honey; recipes for cracklings, pork skins, tenderloin, tails, and feet are also detailed. Apicius' "The Quadruped" contains sixteen recipes for pigs and recipes for suckling pigs alone. Intriguingly, according to the WHO, pigs are, among domesticated animals, rather uniquely "susceptible to the Marburg and Ebola viruses" and can shed them. Pigs are thus a threat in modern settings of filovirus presence as "potential amplifier hosts." The WHO, in fact, has a list of guidelines for pig cultivation in such circumstances to minimize the risk of such species-jumping bat-swine and swine-human zoonotic transmissions. In a third-century rural imperial setting, the initial transmission would have to have been effectively reverse zoonotic with infected humans (possibly having fled from urban Plague of Cyprian outbreaks) first transmitting the virus to susceptible pigs. The swine would then have zoonotically re-infected the rural Romans handling, feeding, and slaughtering them, but not those who merely ate the cooked pork. The situation would have thus been epidemiologically parallel to that noted in Part II for remote jungle logging camps in modern-day Sub-Saharan Africa. "Bushmeat" hunters and butchers are at high risk for EBOV and MARV, but consumption of sufficiently cooked meat *doesn't* result in infection. Such a rather speculative pathway for Plague of Cyprian human rural re-infection via pigs, however, was also dealt a blow, in 2021, by the publication of two new serological studies that cast at least some doubt on the practical existence of such a human–porcine transmission pathway (as opposed to merely theoretically) in modern Sub-Saharan Africa [27, 15, (35.6); 4, 22, pp. 108–141;

37, 14]. Such an eventuality indeed seems to offer only a narrower inferential epidemiological possibility of sustaining an outbreak (among those directly handling the pigs) but not building it to the cataclysmic levels seen in urban areas.

Our "guardrail" modeled outbreak possibilities for rural areas thus seem, when holistically evaluated, significantly less epidemiologically threatening than those for the cities. Nor was the modern overlay of data from the experiences of modern rural Sub-Saharan MARV (and we also considered EBOV) outbreaks particularly countervailing. The outcomes of these modern MARV/EBOV outbreaks ranged very widely—from zero, or merely a few deaths, to hundreds, to thousands or, in one case, tens of thousands. While the lack of modern medical intervention would presumably act as an effective strong R0 enhancer, the far lower population densities (as opposed to the comparatively more densely populated ancient cities) and much more limited means of human movement in the rural areas of the third-century Roman Empire as compared to modern Sub-Saharan Africa would also act as fundamental epidemiological suppressors. In the end, the application of the overall modern scientific evidentiary overlay led to our epidemiological rural model for the Plague of Cyprian, which casts it as far more intensively urban than rural. Yet, the application of other modern evidence, such as for EBOV-related *ville morte* events (detailed in Part I), lead us partially back in the opposite direction, at least in a short-term sense, of R0 enhancement—but only in rural areas in fairly close proximity to afflicted cities. The cumulative impact of ancient pandemic "frictions" would have likely limited (if not eliminated) broader spread in the vast majority of these cases. Violence could have been a contributing factor as well, but as noted many times, it was neither a temporal nor geographic "given" by any means during the Third-Century Crisis. While clearly the Plague of Cyprian spread to (we modeled it doing so, particularly in the vicinity of larger urban areas affected by *ville morte* events or political or military violence) and somewhat impacted rural areas, we remained far more skeptical that this was likely to any great extent. This was true, particularly when compared to our more detailed and clear-cut guardrail epidemiological modeling architecture for third-century urban areas. Mathematical calculations by our team, based on comparative data (2021) drawn from twenty-first century Sub-Saharan Africa, roughly estimated 350% greater overall population densities per square kilometer, for example, in modern times than during the Roman imperial second to third centuries. Such estimates, however, contain many assumptions, uncertainties, and approximations. Comparative overall levels of urbanization are unknown. An ancient city like Carthage, though, was likely even more densely populated (by just under a factor of two) than, for instance, modern Kampala. This would suggest a collective further downward effect of unknown magnitude in terms of modeled R0 values for ancient Roman rural settings. Then, one has to balance the epidemiological effects of modern technological mobility against the lack of modern medical intervention in an ancient epidemiological setting. Holistically and additionally, we cautiously modeled comparatively lower R0 values, through reduction of contacts, for the ancient Roman as opposed to modern Sub-Saharan African rural settings [1, 10, 13, 14, 25, 45].

Our guardrail epidemiological modeling was thus, on the whole, indicative of a more limited rural impact from the Plague of Cyprian (cutting explicitly against what we increasingly believed to have been exaggerated, unreliable, biased, misinformed, or hyperbolic general assessments in certain ancient literary sources). Our default procedure throughout our research was to generally believe the ancient sources unless the systems-synthesis evidence, holistically strongly indicated otherwise. This appeared, on the surface, to be just such an instance. But conversely, we initially did not see our modeling as strongly definitive due to a frustrating lack of direct ancient evidentiary support/confirmation for our systems-synthesis modeling overlay. More conventional literary evidence for rural areas in this regard is nearly nonexistent. It is fundamentally very difficult in an analytical sense to "prove a negative" under such circumstances, and we frankly despaired of ever doing so. In an analytical/evidentiary sense, we were saved from this conundrum by the erudite papyrologically based research that appeared in Dr. Sabine Huebner's 2021 article. Huebner makes a viable case for a lack of a major rural plague impact in the relevant decades in Roman Egypt outside of "the densely populated quarters of Alexandria." There was no rural equivalent to the reference in the papyri to the mentioned "great dying at Alexandria (though that papyrological reference itself, while likely referencing plague, is a bit uncertain as to exact cause or mix of causes)." She also finds no evidence of any "mass mortality event… in the Egyptian hinterland." Through analysis of the evidence of papyri from the Heroninos Archive and other sources, Huebner's scholarship gave us something that we couldn't have accomplished on our own, increased analytical and evidentiary confidence in the implicit overall verdict of our modeling—regarding the limited impact of the Plague of Cyprian on the rural areas of the third-century Roman Empire. Huebner makes a strong case that the papyrological evidence concerning the third century is in fact equal or superior to that of the two preceding centuries in quantity and quality. "We can therefore hardly speak of a dark century; at least at least as far as Middle and Upper Egypt are concerned." Though, once again there are cautions. Such evidence is, by Huebner's implication at least, less plentiful for Lower Egypt, which would have included rural areas in the vicinity of Alexandria where our modeling predicted higher levels of epidemic infection. Matters of exact dating, as usual, also prove consistently troublesome in terms of interpreting the available papyrological evidence. This is all, however, very clearly and evenhandedly explained by Huebner [24, pp. 165–168].

Not that the plague was absent altogether. Huebner found potential evidence of rural cases of the Plague of Cyprian "in at least two papyri dating to the third century." There is also a highly intriguing letter from "a certain Lykarion to his father." This letter references a seemingly large and deadly pestilence. Its provenance and dating, however, are quite uncertain. We do note the aforementioned issue of missing or unprocessed/untranslated papyri and also that none of the papyri seem assignable in provenance to Luxor (or for that matter Myos Hormos) or the areas immediately around either. On the whole though, Huebner argues convincingly that there simply should have been more available if the plague had impacted rural areas heavily (as it clearly did Alexandria). Overall, the verdict of our rural modeling assumptions thus finds some strong support and no contraindication. Huebner also notes, on the basis of papyrological evidence paleographically dated to the third century, that at least one important regional urban node (Oxyrhynchus) was, potentially at least, heavily impacted by the Plague of Cyprian (Kyle Harper had also so concluded and we concur). This is on the basis of a letter from a certain Pausanias to his brother Heracleides. The letter is interesting for potentially documenting an additional major Plague of Cyprian urban outbreak and for its revelatory quality in terms of Roman epidemic disease awareness and avoidance behaviors. Huebner's overall interpretation of that evidence regarding a potential Plague of Cyprian outbreak at Oxyrhynchus is, however, somewhat more skeptical than Harper's (and ours). Our concurrence with Harper's view is due to a perception that Huebner's skepticism seems rooted in certain exclusionary epidemiological assumptions that are overly simplistic (we reference our previous discussion of R0). By contrast, Arsinoe, a significant city and regional administrative center in Roman Egypt does not seem to have been, in Huebner's convincing analysis, heavily affected at all. The tentative conclusion of our, very guardrail in this instance, modeling had been that these much smaller urban centers, with populations under 30,000 people, slightly off the main trade and movement routes, but positioned near to them would occupy complex and unpredictable "gray areas" influenced by a stochastic and difficult to evaluate epidemiological modeling mix of infectivity-retarding ancient movement frictions counterbalanced by some of the infectivity-enhancing urban practices of the larger cities. These offsetting factors would be operating, however, among smaller and presumably less spatially dense urban populations. While that larger analytical matter is definitely complex and uncertain, such an overall interpretation seems to find at least tentative support in the papyri [24, pp. 165–168; 21, p. 235].

References

1 Ajelli, M, Merler, S, 2012 "Transmission Potential and Design of Adequate Control Measures for Marburg Hemorrhagic Fever," *PLoS One*.https://doi.org/10.1371/journal.pone.0050948.

2 Allaranga, Yokouide and Kone, Mamodou Lamine, et al., 2010 "Lessons Learned During Active Epidemiological Surveillance of Ebola and Marburg Viral Hemorrhagic Fever Epidemics in Africa," *East African Journal of Public Health* (1): pp. 30–6 *PMID*: 21413569. https://www.pubmed.ncbi.nlm.nih.gov. Accessed on 8/12/2021.

3 Aparaschivei, Dan, 2012 *Healthcare and Medicine in Moesia Inferior Colection Bibliotheca Iassiensis* (Bucharest).

4 Apicius, *De Re Coquinaria* (inclusive) . https://www.knowtheromans.com. Accessed on 12/04/2020.

5 *Bava Mitzia* 107b; *Hullin* 24b. *The Jerusalem Talmud* 2010.

6 Bodel, John, 1999 "Death on Display: Looking at Roman Funerals," in Bergman, B. and Kondoleon, C., eds., *The Art of Ancient Spectacle* (Washington): pp. 258–281.

7 Bodel, John, 2004 "The Organization of the Funerary Trade at Puteoli and Cumae," in Panciera, S., ed. Libitina e dintorni (Libitina 3) Rome): pp. 149–170.

8 Borchert, Matthias, Mulanga, Subue, et al., 2006 "Serosurvey on Household Contacts of Hemorrhagic Fever Patients," *Emerging Infectious*

Diseases Journal. https://wwwnc.cdc.gov/eid/article/12/3/05-0622_article. Accessed on 3/18/2022.
9 Cartwright, Mark, 2013 "Roman Baths." https://www.ancient.eu/Roman_Baths/. Accessed on 8/14/2021.
10 CDC, 2021 "History of Ebola Outbreaks." https://www.cdc.gov. Accessed on 7/23/2021.
11 2008 "Death and Bereavement in Judaism: Ancient Burial Practices," https://www.jewishvirtuallibrary. Accessed on 12/13/2021.
12 Dvorjetski, Esti, 2017 *Leisure, Pleasure, and Healing: Spa Culture and Medicine in the Ancient Eastern Mediterranean* (Brill, Leiden)
13 2018 "Ebolavirus and Marburgvirus Infections," *The Center for Food Security & Public Health*, July 2021. https://dr.lib.iastate.edu/handle/20.500.12876/14145. Accessed on 7/19/2022.
14 Eliav, Yaron, 2000 "The Roman Baths as a Jewish Institution," *Journal for the Study of Judaism* 31(4): 416–454.
15 *Epitome De Caesaribus*: A Booklet about the Stile and Manners of the Imperators Sometimes attributed to Sextus Aurelius Victor Trans. Thomas Banchich 2018, 35–36. https://www.roman-emperors.sites.luc.edu. Accessed on 5/4/2022.
16 Eusebius Pamphilus, *Ecclesiastical History* Trans. A.C. McGiffert, 1890 Pantianos Classics.
17 Gilad, Elon, 2015, April 21 "The History of Jewish Burial Rites," *Haaretz*, https://www.haaretz.com/.premium-the-history-of-jewish-burial-rites-1.5353617. Accessed on 12/18/2021.
18 Gill, N.S., 2019 "Roman Baths and Hygiene," https://www.thoughtco.com/hygiene-in-ancient-rome-and-baths-119136. Accessed on 10/2/2020.
19 Hanson, Ann, 2006 "Roman Medicine," *A Companion to the Roman Empire* (Blackwell, New York).
20 Hanson, J.W. and Ortman, S.G., 2017 "A Systematic Method for Estimating the Populations of Greek and Roman Settlements," *Journal of Roman Archaeology* 30: 301–324.
21 Harper, Kyle, 2015 "Pandemics and Passages to Late antiquity: Rethinking the Plague of Cyprian c. 249–270," *Journal of Roman Archeology* 28: 223–260.
22 Harper, Kyle, 2017 *The Fate of Rome: Climate, Disease & the End of an Empire* (Princeton University Press).
23 *Historia Augusta, LacusCurtius* Trans. David Magie 1932. https://www.penelope.uchicago.edu. First accessed on 4/20/2020.
24 Huebner, Sabine, 2021, June "The 'Plague of Cyprian': A Revised View of the Origin and Spread of a 3rd-c. CE Pandemic." Published online by Cambridge University Journal *of Roman Archaeology* 34(1): 151–174.

25 Kelly, John Daniel, Wannier, Sarah Rae, et al., 2020 "The Impact of Different Types of Violence on Ebola Virus Transmission During the 2018-2020 Outbreak in the Democratic Republic of Congo," *Journal of Infectious Diseases* Vol. 222.https://doi.org/10.1093/infdis/jiaa163. Accessed on 3/18/2021.
26 Killgore, Kristina, 2016 "6 Practical Ways Romans Used Human Urine and Feces in Daily Life," https://www.mentalfloss.com/article/76994/6-practical-ways-romans-used-human-urine-and-feces-daily-life. Accessed on 2/2/2021.
27 "Livestock in the Roman Empire: Mostly Pigs and Sheep," 2016 *Facts and Details*. https://www.factsandetails.com. Accessed on 9/20/2020.
28 McCane, Byron, 2022 "Burial Practices in First Century Palestine," https://www.bibleodyssey.org. Accessed on 8/14/2022.
29 Mehedi, Marfique, Grosseth Allison, et al., 2011 "Clinical Aspects of Marburg Hemorrhagic Fever," https://pubmed.ncbi.nlm.nih.gov/22046196/. Accessed on 2/3/2021.
30 Miller, S.S., 1992, "R. Hanina bar Hama at Sepphoris," in Levine, L.I., ed., *The Galilee in Late antiquity* (New York): pp. 175–200.
31 Mingoia, Jessica, 2022 "Baths of Caracalla," https://www.smarthistory.org. Accessed on 7/19/22.
32 Moseley, Angela, 2021 "The Art of Christian Burial in Late Antiquity." https://www.dokumen.tips. Accessed on 8/4/2021.
33 Nenova, Stella 2016 "Ancient Greek and Roman Bathing," https://www.medium.com. Accessed on 11/12/2021.,
34 Nock, A.D., 1932 "Cremation and Burial in the Roman Empire," *The Harvard Theological Review* 25(4): 321–359.https://doi.org/10.1017/S0017816000021313. Accessed on 7/11/2022.
35 Paulus Orosius, 1964 *The Seven Books of History Against the Pagans* Trans. Roy Deferrari. Catholic University of America Press (Washington).
36 Preston, Richard, 1995 *The Hot Zone* (Anchor).
37 2018. "Reverse Zoonosis: Human Pathogens are Ruining Wildlife," https://www.healtheuropa.com. Accessed on 9/18/2020.
38 Seneca, 2003 "Living Above Roman Baths," in Lewis, Jon, ed., *The Mammoth Book of Eyewitness Rome* (Carroll and Graff): p. 206.
39 Shultz, James, Cooper, Janice, et al., 2016 "The Role of Fear-Related Behaviors in the 2013–2016 Ebola Virus Disease Outbreak," *Curr Psychiatry Rep.* https://doi.org/10.1007/s11920-016-0741-y. Accessed on 4/6/2021.
40 Slenczka, Werner and Klenk, Hans Dieter, 2007, "Forty Years of Marburg Virus," *The Journal of Infectious Diseases* 196(Supplement 2): S131–S135.
41 Tomorad, Mladen, 2012 "Ancient Egyptian Funerary Practices from the First Millennium BC to the Arab Conquest of Egypt" (c. 1069 BC- 642 AD) *The Heritage of Egypt* 2: pp. 10–40.

42 Toynbee, J.M., 1971 *Death and Burial in the Roman World* (Johns Hopkins University Press): pp. 43–45.
43 WHO, 2021 "Fact Sheet' Marburg Virus Disease." https://www.who.int/news-room/fact-sheets/detail/marburg-virus-disease. Accessed on 3/19/2022.
44 Williams, Yona, 2009 "Massage Therapy in Ancient Greece and Rome." https://www.unexplainable.net. Accessed on 10/20/2020.
45 Embedded comparative calculation based on the following sources... World Bank 2021 Modern Sub-Saharan African Population density (people per square kilometer of land area https://www.data.worldbank.org. The World Bank Accessed on 2/7/2021. Also, Frier, Bruce W., 2000 "Demography," in Bowman, A. Garnsey and Rathbone, D., eds., *The Cambridge Ancient History,* (Cambridge University Press): pp. 787–816. [To match the World Bank data, we rounded up Frier's calculations to the nearest whole number]. See also Hanson, J.W. and Ortman, S.G., 2017 "A Systematic Method for Estimating the Populations of Greek and Roman Settlements," *Journal of Roman Archaeology* 30: pp. 301–324, p. 319 and 2015, "Kampala" *Atlas of Urban Expansion.* https://www.atlasofurbanexpansion.org. Accessed on 8/18/2022.
46 *Yerushalmi Shab* 14b *The Jerusalem Talmud* 2010.
47 Zeldovich, Lena, 2021 "How the Ancient Romans Went to the Bathroom," *Smithsonian Magazine/RealClearScience.* https://www.realclearscience.com. Accessed on 1/8/2022.
48 Zosimus, 2004 *The New History*, Taylor Anderson, ed. (Odin's Library Classics).

CHAPTER 9

The Plague of Cyprian: Timelines, Outlines, and Parameters

The Fragmentary Chronology of a Pestilence, 251–270 CE (Part 1)

The Plague of Cyprian spread throughout much of the Roman World, and to Rome itself within roughly two and a half to four years (248 to 252 CE). A comparison to the second century Antonine Plague, which spread somewhat faster from Asia Minor to Rome and beyond between 165 and 166 CE, is interesting. If that earlier pestilence was indeed caused by the fairly weakly aerosolized (infection range 3–4 feet) pathogen smallpox, as is widely surmised, the slower spread would (assuming general socio-environmental Ro/Re influencing factors between the second and third century to be fairly similar) again suggest a "slower" non-aerosolized pathogen such as MARV as the Plague of Cyprian pathogen (as opposed to, for instance, a pandemic caused by an aerosolized influenza virus). After all, aerobiology of infectious disease creates "a much higher number of susceptible individuals [potentially increasing R0 values] at a much greater distance from the source of the infection"; though admittedly, the complex and variable vagaries of R values make this less than certain in all instances. The Plague of Cyprian also seems to have sustained an outbreak that lasted over a decade—15 years according to the account of the Athenian historian Philostratus—as likely transmitted (yet again) through the later Byzantine account of Evagrius Scholasticus. This timeline is also echoed, in a somewhat confusing way, by Orosius. Such a description

M. Orsag et al., *Interdisciplinary Insights from the Plague of Cyprian*,
https://doi.org/10.1007/978-3-031-26094-0_9

is also bolstered by evidence from *Historia Augusta*. Both the accounts of Orosius, who noted an "unusually continuous and severe pestilence," and also more vaguely of Zosimus, when his treatment of the relevant events is treated holistically, also bolster the case for conceptualizing the Plague of Cyprian as a single, but multi-wave geographically vast outbreak of the "same disease" centered in, if not exclusive to, urban centers as opposed to separate events, geographically or temporally, potentially involving multiple pathogens. Nothing about this broader timeline seems a poor fit with MARV. It has sustained multi-year modern outbreaks in Africa despite modern medical intervention. Our research and analysis, differing only slightly from Kyle Harper's, postulates a roughly fifteen-year outbreak duration. Though far from a major point, this matches the fifteen-year timeline attributed to the ancient historian Philostratus [17, pp. 231–235; 15, 19 (The Lives of the Two Gallieni, 5.5); 23, pp. 315–316 (7.21–22); 32, pp. 9–11 (1.16–21)].

Dr. Sabine Huebner's alternate theory of the Plague of Cyprian's origins dates the Plague's outbreak to the late summer/early Fall of 251 CE and the depredations of the "Scythian"—actually Goth (as well as Borani, Urugundi, and Carpi) barbarian invaders on the Roman Empire's eastern frontier in Europe. Though no ancient source makes a direct association of the "Scythians" with the pestilence, sources as diverse in perspective as Orosius and Zosimus render the two events as generally contemporaneous. Huebner also cites *Oracula Sibyllina,* the fourth-century historian Aurelius Victor, etc. From our systems-synthesis perspective, however, we questioned (and ultimately rejected) Huebner's origin and outbreak dating hypotheses for a number of reasons. Some of the evidence presented was vague and/or open to multiple plausible interpretations. One passage from Orosius cited by Huebner could very plausibly be interpreted in precisely the opposite way to which Huebner interprets it. Orosius' noted that "where the edicts of Decius for the destruction of churches circulated, to these places a pestilence of incredible diseases extended." To us, this would seem to actually date the outbreak of the Plague of Cyprian to late 249 to mid-251 CE (at the latest)—essentially matching Harper's original dating and earlier than Huebner's mid to late 251 CE outbreak dating, which she links strongly to the death of Hostilian. This is also a claim that Orosius, albeit a hardly unbiased source in this regard, makes quite explicitly as well. "To pass over earlier persecutions, when the persecution of Christians was inflicted by Decius, a great pestilence harassed the entire Roman Empire" [20, pp. 156–160; 23, pp. 315–316 (7.21–22)].

As noted, the cause of Hostilian's death (whether from plague at Viminacium or at Rome by assassination at the hands of his imperial colleague Gallus) is disputed in the ancient sources. Huebner's analysis is, however, impeccably transparent and even-handed in dealing with this evidentiary complication. Orosius also describes an empire-wide pestilence seemingly throughout the reign (251–253 CE) of Gallus and Volusianus—"famous for this plague alone." If Orosius' "great pestilence [that] harassed the entire Roman Empire" only broke out after being brought by the "Scythians" fairly late in 251 CE, epidemiologically, this seems an extremely fast speed of transmission for the ancient world [23, pp. 315–316 (7.21–22)]. This brings us to what we see as the most fundamental and highly problematic conflation in Huebner's origin and timeline hypothesis. Her analysis seems, overall, to nearly impossibly temporally "telescope" the initial plague outbreak (which would have been regional and limited) with the full-blown following "first wave" pandemic phase of the Plague of Cyprian—with both apparently occurring in late 251 CE. An extremely infectious rapidly spreading aerosolized pathogen (McNeill's alternate implication of measles comes to mind) might be (we didn't attempt to epidemiologically model it in a guardrail or any other sense) able to theoretically account for such an accelerated timeline, but this possibility simply cuts against so much of the available evidence, in terms of epidemiology and pathology, that we have for the Plague of Cyprian. There are also no known lineages of aerosolized pathogens that fit the profile in terms of pathology, in many ways, particularly in terms of the extremely high CFR. A non-aerosolized pestilence would have to have originated more narrowly and regionally and then spread more gradually until it was more broadly present—a far from instantaneous progression subject to all Harper's aforementioned epidemiological "frictions" of the ancient world. A less expensive journey from Aquincum to Rome begun in September 251 CE, for example, would have taken nearly two months. Such a direct epidemiological pathway seems, in context, highly unlikely for obvious reasons. Cascading outbreaks along the way, however, would have taken far more time to develop and spread. Yet, Aurelius Victor's cited narrative seems to place the death of Hostilian and Gallus and Volusianus winning praise for paying for the funerals of poor plague victims in Rome in direct temporal proximity. Huebner also cites Zosimus in this latter regard; we, however, could find no such specific reference in *The*

New History. Though this discrepancy may or may not be the product of non-matching translations and numerations. From the beginning of this project, we were very sensitive to any potential problems of "translation shopping" and did all we could to scrupulously avoid any such perception. We chose a single translation without having read it beforehand and relied on it throughout. Both Dr. Sabine Huebner and our team both used the same translation of Orosius [23], for example, but the utilized translations of Zosimus differed [22, 20, pp. 156–160 and 172–174; 32, pp. 8–15 (1.16–21)].

At the heart of Huebner's origin theory, there also lie other contradictions and conundrums. If the disease was largely confined to urban areas in terms of major impact (as Huebner herself concludes and we concur), how did the non-urbanized Goths and other Germanic peoples spread it so widely? Furthermore, if the Goths were themselves suffering from the deadly plague as they attacked the imperial frontier, why was their military performance not more negatively impacted—as that of the Romans appeared to be later against the Sassanids? Instead, the victory at Abritus and the killing of a Roman emperor in battle were, in many ways, utterly unprecedented successes against the traditionally militarily superior Romans. In fact, a reversal of such a Goth-Roman disease analytical "assignment" seems far more logical in context.

When considered holistically, broader geographical realities and likely patterns of movement at the time also seem to cast doubt upon, and indeed provide more probable alternatives to, Huebner's implication of a Balkan origin for the Plague of Cyprian. Our own research, however, fully supports her conclusion that the Balkan region was indeed hit very hard by the Plague of Cyprian in the early 250s CE. The difference lies in why. We contend that the plague was in fact far more likely brought to the region, not by the barbarian invaders themselves, but rather by the heavy influx of men and material from the rest of the Roman Empire meant to drive the invading Goths and other tribes out. The Danube, ***the only river in Europe that flows west to east and north to south*** would have been a convenient main artery in such ***an inflow***. The directional differences in terms of volumes, travel times, and the implications for speed and ease of epidemiological transmission are stark. Going with the Danubian flow from Boiodurum to Aquincum in September 251 CE, for example, would have taken less than five days. Going in the opposite direction would have taken more than eleven. The inflow of men and material to fight the barbarians would have been both intensive and extensive. Soldiers would have arrived by sea, by river, and by land, in the quickest manner possible. Such procedures were well established as the Danube frontier had long been an imperial trouble spot. Egypt and other areas in the Middle East would also have supplied recruits as well as existing forces. There is an applicable, in context, letter from an Egyptian military recruit in the late second century named Apion, who was assigned the Romanized "military name" Antonius Maximus and eventually stationed in Italy, describing this very transportation and induction process in some detail. By contrast, who would have been going in the other, geographically less favorable direction? Much commerce would have been disrupted. The victorious barbarians eventually struck a favorable deal with Gallus to accept tribute and return to their own territory. This was an agreement, however, that they don't seem to have honored for long (launching another regional incursion into Asia Minor and Moesia in 253 CE until driven back). In light of this, Roman military redeployment out of the region would have likely been limited. A large military movement into the heart of the empire that could have served as a more rapid and extensive epidemiological driver doesn't seem to have occurred until mid-253 CE, associated with the successive Civil Wars of Aemilianus and Valerian—too late to support Huebner's dissemination theories [22, 2, p. 374; 32, pp. 8–9 (1.16–17)].

As we'll see later in Part III, there is good archaeological and numismatic evidence of a considerable flow of refugees out of the Danube Bend region in 251 CE. Much of their travel, however, would likely have been slow, short range and overland as they would have lacked financial means. How a plague would have spread so unusually rapidly in 251 CE, even with an aerosolized pathogen, is thus unclear, to say the least. There is more. The Plague of Cyprian was described in the ancient sources as a unique and unprecedented event, nor does anything like it seem to have been repeated. Yet Romans and Goths had and would interact fairly continuously in the coming centuries. Never had nor would again "so great a mortality" (in the words of Zosimus) envelop the two ancient civilizations [32, pp. 8–9 (1.16–17)]. Why? Also, implicitly central to Huebner's analysis is her conclusion that the Plague of Cyprian heavily impacted "... densely settled cities or cramped military camps." This is indeed how the non-urbanized "Scythians" could have plausibly, and analytically consistently, spread the plague in an epidemiological sense. While the former is clear, however, the latter is quite uncertain. The Roman military environment, in terms of both epidemiology and potential progression of pathology (due to the prevailing high standard of palliative care) was, in context, a setting unto itself. We hope to model and examine that environment specifically ourselves from a systems-synthesis perspective in future research. We are thus in no firm position to argue that Huebner is incorrect in this inference but rather that the limited evidence that she presents for it in the 2021 *Journal of Roman Archaeology* article was, at least to us, not analytically persuasive. That evidence derives from the accounts of Zosimus and the sixth century Eastern Roman official and historian Peter the Patrician concerning Valerian's military campaigns against the Sassanid Persians in 259–260 CE. The brief and fragmentary account of Peter, as well as Thomas Banchich's leading scholarly interpretation of it, implicates the city of Edessa as a potential epidemiological driver of a plague outbreak that damaged the Roman military. In a series of events "set at or near Edessa" in 260 CE, Valerian is depicted as collecting "an untold amount of gold" to try to pay off the invading Persians under King Sapor I. This after "plague struck his [Valerian's] bases," unnerving the emperor. Units of the key Moorish cavalry (who would play a key role in Aurelian's later victories) were particularly affected. One of those bases, and the likely center for Valerian's collection and storage of the tribute payment, was the city of Edessa. These developments led to one of the most cataclysmic military defeats ever suffered by the Romans, the Battle of Edessa and the capture of Valerian by the Persians [20, pp. 161–162 and 170; 3, pp. 113–114)]. Zosimus' account of the events of 259 CE is even clearer in this same regard:

Valerianus had by this time heard of the disturbances in Bithynia, but his district would not allow him to confide the defense of it to any of his generals. He therefore sent Felix to Byzantium, and went in person from Antioch into Cappadocia, and after he had done some injury to every city by which he passed, he returned homeward. But the plague then attacked his troops.... [32, pp. 10 (1.19–21)].

Valerian's undisciplined legions had clearly entered and looted multiple "densely settled" imperial cities. Only after carrying out those destructive multiple urban forays had plague struck the troops. When considered in an epidemiological context, Huebner's implicit conclusion that the third century Plague of Cyprian thus provably disseminated rapidly specifically in a military barracks or camp environment, like the second century Antonine Plague had, seems unsupported—at least so far; the accounts' joint indirect implication of urban epidemiological environments does not permit, in our view, such a firm inference [20, 18, p. 99].

A final analytical challenge loomed for Huebner's theory of Balkan origins for the Plague of Cyprian when we attempted to deduce what the pathogen could have been in the relevant biogeographical context. Leaving any wholly unevidenced speculation regarding extinct pathogen lineages aside, we have already differentially eliminated many diseases. Considerable evidence points to viral hemorrhagic fever. While Huebner's 2021 article lists the pathogen as "unknown" she too seems, at least implicitly, to acknowledge a strong possibility for a VHF. Our Map 3.1 shows two biogeographically included VHFs. One of these, hantavirus HFRS, we have already eliminated due to its dependence on rodent droppings for transmission, lack of human-to-human transmission and therefore, ability to cause large outbreaks and spread rapidly. It is largely rural in epidemiology and usually displays a too-low CFR [20, pp. 156–160 and 170–171].

The second biogeographically included VHF from Map 3.1, Crimean-Congo Hemorrhagic Fever (CCHF) is initially intriguing and far more viable as a final alternative Plague of Cyprian pathogenic suspect. Pathology and progression, in terms of stages, of the more severe cases are highly included; in fact, very similar to (if still milder than) Ebola or Marburg. Specific elements of deduced modern epidemiology, such as human-to-human transmission centered in nosocomial environments and the infectiousness of corpses find strong ancient parallels in the accounts of Cyprian, Orosius or in Eusebius' *Ecclesiastical History*. The overall CFR for CCHF, however, is far lower than those of the two deadly filoviruses—only around 3.3–5% (this, of course, takes into account modern medical treatment). Nevertheless, the widely endemic disease causes around 500 deaths a year in the twenty-first century. As its name implies, it is a very widely biogeographically distributed pathogen from the Russian steppe to Africa and Asia (including India and China). Its annual outbreak "hotspots" are exactly on target in context—in the Balkans, Eastern Europe, and the Near East. In a molecular-phylogenetic sense, this is also a very slow-evolving and truly ancient pathogen. Its TMRCA dates back to 1500 to 1100 BCE! In an unprecedented, and as yet unreplicated 2017 archaeological/paleo-genomic find dating from 600 to 450 BCE, CCHF viral proteins were extracted, by a team of researchers at Purdue University led by Dr. Conner Wiktorowicz, from ceramic pottery fragments whose provenance was from the Danube region of what is today Germany. Human blood and organs had been, uncharacteristically for the region and time period, buried in ceramic pottery. By grinding up some of the pottery fragments, the Purdue researchers isolated CCHF viral peptide proteins from a decaying film on/in the ceramic pottery. Thus, CCHF was almost certainly biogeographically extant during the Plague of Cyprian in the Danubian region in very much its modern form (though there are a variety of extant strains of this ancient, widespread, and genetically diverse VHF). To close out an open analytical thread from Part I, it is also important to note that these remains *were not* burnt on a pyre as part of *bustum*-style burial practices, contrastingly to what was evidenced for the possible "Tomb of Hostilian" at Viminacium by Dr. Sabine Huebner [5, 10, 11, 12, 16, 20, p. 157; 25, 27].

There is thus no need to refer, in context, to an evolutionary lineage in regard to CCHF. However, upon closer examination, the case for CCHF as the Plague of Cyprian pathogen breaks down. If CCHF was endemic in the Danubian region in the third century CE as seems nearly proven, why the unique flare up into a massive and novel pandemic? This is simply not the epidemiological pattern of CCHF. Its outbreaks show a low-level recurring endemic pattern across its vast biogeographic range. As a virus "generally mediated by a tick vector," some of the same disqualifying factors described earlier for LBRF apply, though perhaps not as strongly, for urban areas. Primary (and indeed secondary) transmission follows a "tick vertebrate (with generally asymptomatic animals functioning as reservoirs and amplifiers)-tick ecological cycle"; thus, displaying a peak June–July seasonality that does not match that of the Plague of Cyprian. CCHF's epidemiology is indeed primarily rural, and it often mainly affects agricultural workers. Slaughterhouse workers are also an occupationally vulnerable population in an epidemiological sense. It is very biogeographically widespread (probably disseminated by ticks on migratory birds). Human-to-human transmission indeed occurs with CCHF, but it is a limited, indeed tertiary, factor in CCHF epidemiology. Up to 90% of CCHF human infections result either from direct tick bites or from exposure to tick-infected animals (whether wild or livestock). The biogeographic range of the virus essentially overlaps with that of the *Hyalomma* tick (though other tick species can host and spread CCHF). Only 1 in 8 infected humans even presents with more severe pathogenesis. As noted, 88% of all human CCHF cases are mild, indeed often subclinical, with an overall CFR of only about 15% in the more severe cases. Documented

Plague of Cyprian locations are also to the north of the 50-degree parallel that marks the biogeographical range limit of CCHF's reservoir host the *Hyalomma* tick. These are areas where modern outbreaks of CCHF don't occur, even in today's mobile and highly interconnected world. This northern latitude thus constitutes a biogeographical barrier that the disease's tertiary human-to-human vector is seemingly not epidemiologically viable enough to supersede. Though climate change (resulting in northward expansion of *Hyalomma's* range) may one day change things, this barrier is so far, inviolate, across the Eurasian landmass—Western Europe, Eastern Europe, Scandinavia, and the Russian Federation (where, not counting annexed Crimea) only the southerly North Caucasus and Pontic-Caspian Steppe regions of Russia record cases of CCHF. This overall broader reality simply does not fit with what we know about the Plague of Cyprian in an epidemiological sense. In the end, the case for CCHF is thus not a convincing one [5, 11, 16, 24, 26, 27].

From the initial African outbreak (from "Ethiopia"/Aksum to Luxor, Alexandria and Carthage) recorded by such sources as the *Excerpta Salmasiana*, Symeon the Logothete, George Kedrenos, John Zonaras, Dionysius/Eusebius, Cyprian, and Pontius, some of which we have already detailed and analyzed, our broader conclusions thus strongly (if not exactly) echo those of Dr. Kyle Harper—who proposed a 249–262 CE timeline for the Plague of Cyprian. Details of the subsequent waves of the Plague of Cyprian after the first few years of the outbreak, however, are often very fragmented and lacking in detail. The multifaceted Third Century Crisis with its civil wars, rampant inflation, the decisive fading of the Roman Climate Optimum (RCO), famines and supply chain breakdowns, de-urbanization, military anarchy, barbarian invasions and eventual de facto political fragmentation of the Roman Empire immeasurably added to the destruction and social, political, economic and military disruption of the plague itself. Yet, we can glean some key additional details from this very limited evidence. Again, Harper's invaluable marshaling (23 distinct literary sources) of the scattered strands of ancient evidence and overall analysis in this regard are highly impressive and informative. The Byzantine writer George the Monk recorded a detail that some cities (apparently including Rome itself) suffered through two waves of the Plague of Cyprian [17, pp. 233–236; 19 (The Lives of the Two Gallieni, 5.5–5.6)]. Ultimately, flight stemming from fear and forms of Roman "rough and ready" disease awareness (along with the presumed at least temporary collapse of the functioning of the urban "super spreader" institutions) would have helped, after an initially counterproductive *ville morte*-driven but localized spread, eventually helped contain the big urban outbreaks. The interrelated collapses of the Third Century Crisis would have probably had more contradictory effects overall. A gradual breakdown in empire-wide travel volumes would have introduced extra "friction." Cooler temperatures associated with the decisive fading of the warmer RCO might have, however, worsened the effects of the Plague of Cyprian given its seasonality. De-urbanization would have vastly diminished the potentially devastating urban R0 enhancers detailed in Table 7.2 over the longer haul, but the violence and disruption of urban conflict, *ville morte* events and barbarian occupation would have been powerful potential short-term dissemination enhancers in and of themselves.

De-urbanization in the shorter term, as noted, would also have facilitated disease dissemination of the plague to the surrounding countryside. From what we know of it from the admittedly patchy evidence, the last massive Gallienic wave of the Plague of Cyprian fits this pattern quite well—including *ville morte* and other massively de-urbanizing events taking place against a background of violence, invasion, civil war, generalized chaos, political fragmentation, and economic breakdown that marked the peak of the Third Century Crisis. A more or less final subsidence of the pestilence seems to have followed that blazing intensification in a manner that broadly, if not deeply, suggests a highly deadly non-aerosolized pathogen linked, in an epidemiological sense, very specifically to both violence and the R0 functioning of third-century urban environments. Though the usual problematic and reductionist circularities and evidentiary vagaries limit the value of such a conclusion, it too "fits" in a way consistent with the other Plague of Cyprian puzzle pieces. An uncontradicted, if far less than definitive, consilience once again seems to exist.

Numismatic and Archaeological Evidence from Brigetio: A Possible Filovirus "Ville Morte Event" Microcosm of the Impact of the Plague of Cyprian?

On the Roman Empire's Danubian frontier (in modern-day Hungary), the civilian town of Brigetio in 250 CE was situated near surrounding rural villages and *cannabae* associated with *a* legionary camp and fort that had long served as the base for *Legio I Adiutrix* along with, more recently, a detachment of *Legio II*. A Roman road that seems chiefly to have connected the area with the provincial capital of Aquincum and the other more "inner" towns of the province of Pannonia Inferior ran to and from Brigetio. The Roman navy's *Flavia Pannonia* also had a base at Brigetio as the area existed at a confluence of land and riverine commercial and military transport links. Along with soldiers, sailors, and their families, the area's population seems to have included many Syrian immigrants, who had relocated there after the second-century Antonine Plague and the Marcomannic Wars had taken a heavy toll on the region's original Italian and Pannonian inhabitants. The town boasted an amphitheater and an aqueduct over 15 kilometers long that brought in the unusually warm spring waters from around Lake Tata. A system of lead and terracotta pipes distributed the warm waters to civilians and soldiers alike. The "thriving" civilian town seems to have been particularly known for the healing properties of these thermal spring waters. An impressively large Roman bath and spa complex of well over 1500 square meters, gradually unearthed by archaeologists since 2014, seems to have effectively dominated the 10,000 square meters ancient civilian town and accompanying cemetery). As at Hammat-Gader in Israel, the local thermal springs are still used in a similar manner today as in Roman times, for relaxation and medical purposes, such as healing rheumatic ailments. Styling itself in the "tradition of spas" that flourished already in Komarom, "in the age of the Romans," the *Brigetio Thermalspa of Komarom* is now situated "in the heart" of the modern Hungarian town [4, 7, 9, 21, 28].

Yet, the ancient ruins present a conundrum concerning some sort of unexplained mid-third-century semi-apocalyptic event concerning this ancient town. In the early 250s CE, apparently sometime between 250 and 253 CE, life in the civil town of Brigetio, with its magnificent, and apparently well-known spa and baths, and its amphitheater came to a fairly "sudden halt." Brigetio was evacuated and became, effectively, a *ville morte*. Numismatic evidence shows that monetary circulation, both of genuine coins and ancient counterfeits, likely terminated during the reign of Emperor Trebonianus Gallus (251–253 CE). Buried hoards in the nearby region also terminate with coins of Trebonianus Gallus. Burials at the cemetery also abruptly ceased around the same time. The population fled, not overnight apparently, but fairly rapidly. Yet, no archaeological evidence of barbarian invasion or occupation exists; "there are no destruction layers in the municipium of Brigetio." Nor do surviving "written sources mention any particular invasion or threat [at that time] on this part of the border." The only sign of any burning in the city was a small fire apparently set by someone passing through *after* the city was evacuated. Other cities in the region, such as Aquincum, also seem to have seen similar exoduses at around the same time, but they later recovered. Until, in Aquincum's case, it was destroyed during a mid-fourth-century Sarmatian invasion. Brigetio, however, apparently never came back and remained a ghost town. Yet, around the *ville morte* of Brigetio, the military camp and fort, the rural settlements, the *cannabae*, and the road to Aquincum all seemed to continue to function, continuously, well into the fourth century [8, 21].

Our team came across the mystery concerning Brigetio and its environs rather late in our research and writing process. We were struck by how remarkably well this seemingly baffling piece of evidence fit with our concepts and "solution" to the Plague of Cyprian puzzle. Brigetio would have been hit harder and earlier in the pandemic due to the relative ease of access to it; it was a prominent riverine transport, commercial, and military hub. This is a place that would have served as a hub for the inflow of men and material, with the current of the Danube, to fight the Goth invasions in the early 250s CE. Fewer ancient pandemic "frictions", to once again use Kyle Harper's highly relevant terminology, would have applied to it. The other key factor, according to our modeling of the POC, would have been the dominant presence of the linked MARV—viral lineage "super exposers"—the medical spa and the baths, specifically in the, probably suddenly overcrowded, civilian town of Brigetio. As "slow burn" turned to "urban build", the population would have left. The sudden halt to burials might indicate some other emergency means of body disposal—perhaps along the lines of what seems to have happened at Luxor? Though this seems somewhat less sure. The surrounding military and rural areas would have been surely hit as well but evidently not as severely—a survivable "slow burn"—particularly when Roman "rough and ready" pandemic reactions would have led these populations to avoid the baths and the medical spa in the nearby town before these institutions and the town that contained them collapsed, through a combination of plague and fear of plague, and became, in our estimation, a filovirus-created *ville morte*. Timing is also everything here as it relates to the question of the ancient *ville morte* event of Brigetio and the Plague of Cyprian. The years 251–253 CE were, after all, the time of "so great a mortality," or of Emperors "Gallus and Volusianus [being] famous for this plague alone" or having a reign "remarkable only for a pestilence," or when the pestilence "blighted the face of the whole earth," or the Roman Empire was "seized by that general pestilence and laid bare" in the words of Zosimus, Orosius and Eutropius, Jordanes, and (once again) Orosius, respectively. Our solution fits the surviving evidence here perfectly. Any alternative explanation seems, to us at least, far less of a match. Due to the extreme lethality, and at times terrifying symptoms of filoviruses, fear is a particularly important factor in outbreaks, helping to create, among other phenomena, the sort of *ville morte* events described in Part I. Often fear, as much as the pathogen itself (whether Marburg in the Democratic Republic of the Congo or Angola or Ebola in western Africa), is the enemy in terms of modern medical interventions. "Fear driven actions may escalate and reach a tipping point..." The civilian town of Brigetio, it would appear, did just that [32, pp. 8–9 (1.16–17); 23, p. 315 (7.21); 14 (9.9); Jordanes as cited in Harper; 18, pp. 138; 21].

The Fragmentary Chronology of a Pestilence, 251–270 CE (Part II)

The initial wave of the Plague of Cyprian indeed seems to have expanded unabated at least throughout 253 CE (during the reigns of Emperor Trebonius Gallus and his son Volusianus and their successor Aemilianus). Hostilian, the younger son and heir of Emperor Trajan Decius (and initially co-Emperor with Trebonianus Gallus after Hostilian's father and older brother perished in battle against the Goths in 251 CE) perhaps died of the plague (though he also may have been murdered by Gallus). A number of ancient sources, including, as noted above, the fourth-century historian Eutropius, attribute "enormous mortality" to the two-year four-month rule of these unfortunate emperors whose combined "reign was remarkable for a pestilence"—as Harper's analysis notes. The later (sixth century) Syrian chronicle of the Christian theologian Pseudo-Dionysius (also cited by Harper) may also date to this phase of the pestilence and records "giant pits" to dispose of plague victims in the Middle Eastern regions of the Roman Empire. Numismatic evidence furnishes another likely sign of the disruption caused by this massive initial wave of the Plague of Cyprian in Rome itself. Between late 251 and 253 CE, the Roman Central Mint seems to have been severely disrupted; it failed to produce *antoniniani* in sufficient numbers. To compensate, provincial mints began to re-mint *denarii* of Gordian III that had begun to fall out of circulation. The Plague of Cyprian seems the most likely cause of such disruption; additionally, the 251–253 CE timing fits perfectly with the first wave of the disease that engulfed Rome. While such shortages and resulting compensations by provincial mints weren't unknown and other explanations are possible, this temporally focused and highly notable shortfall contrasts interestingly with the more typical and lower profile and more long-term shortage of specific lesser value "small bronze coins" that had prevailed regionally in the Danubian provinces for much of the 230s and 240s CE [14 (9.9), 17, pp. 236–241 and 226; 23, pp. 315–316 (7.21–22); 32, pp. 8–8 (1.16–17); 30, 31].

Two of the apparently more severe later outbreaks are described in Zosimus' fifth–sixth-century *New History*. In Zosimus' account, these outbreaks of the plague seem, once again, potentially linked to war, violence, and disorder (see line 13 of Table 7.2). The association is weaker in the first case and stronger in the second. While many other diseases, such as louse-borne relapsing fever (discussed in Part I), also would strongly fit with such an association, they work less well with the idea of a phased but unitary plague event clearly noted in a range of the applicable ancient sources encompassing such contrasting perspectives as those of Zosimus and Orosius. Orosius describes "an unusually continuous and severe pestilence" that may have lasted as long as fifteen years (the latter wording appears somewhat ambiguous). We detailed some of the reasons for such associations of violence with R0 enhancement for members of the *filoviridae* back in Part I. As noted, an outbreak in 260 CE (though details are somewhat scant overall) seems to have negatively impacted the army ("…the plague attacked his troops") of Emperor Valerian—contributing to his subsequent defeat (one of the most humiliating ever suffered by the proud Roman Empire) and capture by the Persians. A later outbreak under Gallienus in 262 CE is discussed in somewhat greater detail by Zosimus. Goths, Carpi and other invaders had violently occupied Roman cities in Greece such as Nicomedia and Nicaea and even breached the borders of Italy. Yet, the "plague happened to break out in several of the towns more dreadful than any that had preceded it." While one must again be aware of potential hyperbole, the plague seems to have hit the Roman population first and then the barbarians who became "infested with the pestilence" and had to withdraw, along with what was left of the afflicted Roman population, in mutual flight, from the cities the barbarians had occupied. This would have probably resulted in, by now familiar, *ville morte* events in both of these metropoles. This account is notable and also consistent with the concept of a by-then-endemic Plague of Cyprian pathogen (as well as with line 13 in Table 7.2) as it seems to imply spread in a violence-afflicted (and thus temporarily R0-enhanced) urban setting *from* the Roman population *to* the invaders. The fact that the plague was consistently a factor in the continuing broader violent chaos but did not become effectively all-consuming or preventative of it can be seen as another telling point in favor of a very deadly yet modestly contagious pathogen [22, pp. 316 (7.22); 32, pp. 8–11 (1.16–21). Once again, our AHA co-panelist Baylee Staufenbiel's broader insights were helpful here].

Again, the epidemiological mechanics here aren't discernible. Violence may have been a factor, but it also isn't clear how disrupted the urban MARV "super exposure" generating institutions such as the medical system and the baths were prior to, or perhaps even during, the barbarian occupation? Though one may assume some considerable disruption during the latter stage at least. Additionally, in neither Nicomedia nor Nicaea did the combination of plague and the violence and destruction associated with barbarian invasion and occupation result in permanent or even semi-permanent *ville morte* events. Both cities recovered strongly as the empire's economic center of gravity shifted in their direction in the latter third and early fourth centuries. While the Plague of Cyprian was clearly a major factor in specific local outcomes during the Third Century Crisis, it did not operate in a vacuum; economic, military, and political factors also all synergistically played important roles.

Why Did the Plague of Cyprian End in the Early 260 s CE?

Here, our modeling is largely unsupported by any real overlay of ancient evidence and thus becomes more solely reliant on modern medical evidence and/or informed inference. Why, after fifteen years according to Philostratus, did the Plague of Cyprian pathogen disappear and apparently never resurface? There seem to be four apparent possibilities: (1) Pathogen mutation—the MARV lineage evolved into a far less virulent form ending the pandemic. This is certainly possible. Such a MARV lineage outbreak on this apparent scale and duration of around 15 years is certainly utterly unprecedented in the modern world; a mutation or mutations certainly could have arisen from an epidemic of this size. Such an outbreak would easily eclipse even the massive 2014–2016 West

African Ebola outbreak. Yet, on the whole, evidence derived from modern virology somewhat negates this theory. Genomic research carried out in the wake of the year-long Uige MARV outbreak, which involved almost solely human-to-human transmission, by a team led by Dr. Jonathan Towner of the CDC found only "a 7% nucleotide divergence"—essentially very little mutation. The 1998–2000 Durba-Watsa multi-strain Marburg viruses in the two-year DRC outbreak, however, showed greater levels of divergence—around 21%. Yet, that reflected not mutation in infected humans, but rather multiple injections from the zoonotic pool—and thus likely reflected thousands of years of mutation/evolution in the ERB host populations. While MARV does display genetic diversity across its vast geographic range, the divergent viruses, crucially in context, remain fundamentally similar in their pathology and epidemiology. Additionally, there is simply no evidence for the evolved form of the pathogen; though isolated from its biogeographical sphere, it could have died out. Though again, this is wholly speculative. Overall, this explanation seems possible but unlikely—at least as a unitary explanation [29].

(2) A destructive European extreme cold weather event regarding this tropical pathogen lineage can be even more firmly ruled out. Filoviruses can survive preservation with dry ice at temperatures reaching -70 or -80 degrees Celsius. One of our team's authors regularly thus preserves samples that may contain them. Colder temperatures actually improve survival times for viruses both on the carcasses of experimentally infected primates and as viral particles called fomites on surfaces such as door-knobs. (3) "Herd immunity" seems very unlikely at first, but, as with mutation, it too can't be entirely dismissed. Modern serological research on the durability of immunity to MARV in survivors is conflicting (though there is some potential evidence that it is less enduring than immunity to Ebola). The sheer lethality and low infectivity of MARV—the combination of low viral positivity rates and high CFRs seem to weigh heavily against "herd immunity." And yet, serology provides only partial answers on the duration of immunity; nor does it measure immune memory cells themselves (nor all of their activities). Given the modeled massive urban outbreaks, some level of herd immunity may actually (after a fourteen-fifteen-year pandemic in which some cities saw multiple large outbreaks) have been a contributing factor to the Plague of Cyprian's end. (4) Our remaining viable hypothesis is that Third-Century Crisis-linked de-urbanization, disruption of travel and trade between as well as within

some urban epidemiological nodes and growing awareness (if not understanding) of some of the epidemiological urban risk factors combined to put an end to the pandemic. The epidemiological effects of urban violence would likely have been short-term epidemic enhancement and dissemination followed by dispersal, flight, *ville morte* events, etc. Which would again have ultimately led to decreased basic reproduction numbers. The worsening and fragmenting if uneven situation in the empire as a whole in the 260s CE, before the advent of restorative rulers such Claudius Gothicus and Aurelian, and the aforementioned example of Plotinus (whose deductions seem to indicate an awareness [but not an understanding of epidemiologically risky types of contacts]) come to mind here. As suited as a highly lethal MARV lineage pathogen would have been to the Roman urban epidemiological environment in the third century, it also would have taken only moderate levels of change to contain an outbreak of a non-aerosolized pathogen with MARV's fairly low basic infectiousness. Modern epidemiological modeling of MARV has suggested epidemiological containment (characterized by the basic reproduction number falling below 1.0) via as little as a 20% reduction in infection contacts. Likely #4, combined with more limited contributions from #3 and possibly #1 helped bring the R0 (across the empire) below 1.0—finally ending the destructive pandemic [1, 6, 11, 29].

In the wake of such a result and without a biogeographically included zoonotic reservoir from which to re-emerge, the biogeographically alien disease would never have become endemic, cyclic, or even sporadic within the territory of the Roman Empire (or thereafter). As an RNA virus, it would have left no direct scientific evidence behind...any trace in its unique and destructive path. Nothing like it seems, as described in the ancient or later sources, to have ever happened again. Once again, the pieces seem to fit. As interpreted in our modeling, the Plague of Cyprian thus was a monstrous and lethal pestilence that endured and devastated due to its inherently virulent pathology and specific epidemiological fit with the third-century Roman Empire's urban environments. Yet, in a way, it was also a surprisingly fragile "more dreadful than any dread" fated to become unrepeated and eventually, in Kyle Harper's words, a largely "forgotten pandemic" [18, p. 136].

As noted in Part I, we accept Huebner's dating of an Alexandrian outbreak to 259 CE but reject the dating of the outbreak detailed by Dionysius in Chapter 22 of the *Ecclesiastical History* to 263 CE. Thus, we would argue, a seven-to-eight-year gap seems to exist in the surviving sources, after 262 CE, before any sort of plague is mentioned again. This next disease outbreak occurred in the wake of the crucial military victory of Emperor Claudius II Gothicus, a ruler whom Zosimus describes as "adorned with every virtue" in 268 or 269 over barbarian Goth invaders at Naissus. Claudius subsequently battled an apparently interrelated group of Heruli pirates in 270 CE. A plague that started among these latter barbarians eventually began to "infest the Romans, many of whom died" after some of the surviving defeated pirates were resettled within the Roman Empire or admitted into the legions, which were, under the combined impacts of plague, war, and generalized disruption, in desperate need of manpower. The most prominent of the Roman plague victims was the emperor himself. The *Augustan History's* account effectively corroborates that of Zosimus in many ways. Claudius is described as possessing "the valor of Trajan, the righteousness of Antoninus [Pius], the self-restraint of Augustus…." The *Augustan History* also offers some additional telling details in relation to the Goth invaders. After their massive defeat at Naissus ("…the fields are hidden beneath their [the barbarians'] bones"), "the survivors of the barbarian tribes… were also so stricken with famine and pestilence that Claudius now scorned to conquer them any further." [32, pp. 13–14 (1.24–25); 19 (The Life of Claudius Gothicus, 8.4–11.3)].

In both accounts, from a systems-synthesis perspective, the implied epidemiological spread (actual or implicitly feared), is thus the opposite of that implied for 262 CE—*from* barbarians *to* Romans. This indicates, on balance, an external as opposed to an endemic origin for the disease that ultimately killed the heroic emperor. Thus, this seems to have been a different unknown pathogen at work beyond the borders of the Roman Empire and likely *not* the Plague of Cyprian. The seven to eight year "silence" between 262 and 270 CE also can very logically be interpreted as inferentially supporting this conclusion. Thus, the last major outbreak of "our plague" seems to have occurred, in a manner entirely consistent with Harper's original chronology, in 262 CE. So, any extension of the Plague of Cyprian outbreak timeline to 270 CE on the basis of Claudius II's death appears to us invalid. Nor is such a late terminus compatible with Philostratus' fifteen-year plague timeline. The victory at Naissus, however, seemed one of the first signs of Roman recovery from the Third Century Crisis. Yet, the empire had been and was being transformed in many subtle ways that went beyond the direct effects of plague, invasion, de-urbanization, and economic breakdown that had acted as triggers. For instance, archaeological evidence from around Naissus itself, now the city of Nis in modern Serbia, shows the increasing use of seemingly primitive "bone and antler objects" in the daily life of Roman residents in the latter half of the third century [13]. Over a fifteen-year outbreak within the empire, the overall mortality and displacement from the Plague of Cyprian, particularly from the larger urban areas and more easily accessible (via maritime or riverine travel) smaller to medium-sized towns, must have been quite high. The potential, and clearly multi-causal upper limits of lethality and de-urbanization are best demonstrated by the over 60% (albeit quite possibly hyperbolic) depopulation/mortality estimate for Alexandria in Part I. More broadly, for a highly lethal pathogen from the evolutionary lineage of MARV, the potentially devastating impacts of third-century Roman urban R- enhancers in Table 7.2 are clear indeed. The "super-exposer" urban institutions of the medical system, the baths, and complex funerary rites would have interacted with this particular pathogen in ways that indeed would have produced truly horrific results in those specific epidemiological environments.

Though the Plague of Cyprian was clearly a truly destructive third-century Roman pandemic, overall mortality for the Roman Empire from the pestilence was simply impossible for us to even begin to estimate more precisely. The potential dampening effects of what were likely far lower rates of infection and ensuing death tolls in the less accessible inland towns and in the rural areas of a (by modern standards) thinly populated empire may well have been epidemiologically significant. Ancient overland travel "frictions" and vastly thinner population densities would have also played a suppressing role. While the pestilence was indeed likely universal, as a number of the ancient sources claim, our modeling of it suggests that its impact may have been far deeper in some settings and far shallower in others. Overall, the relative paucity of extant source material in that respect was simply not sufficient, in our current view at least, to even attempt the task of sorting through and trying to "solve" such demographic complexities at a distance of over 1,700 years. Calculating and accounting for all the potential variables involved seemed impossible. Nor did our team include a demographer. We must admit, that we just don't know. Additionally, distinguishing ex post facto, between pandemic mortality and other forms of mortality would be next to impossible for the chaotic and often poorly documented years of the Third-Century Crisis. Despite disagreeing with her temporal fixing of the pandemic's end date, in terms of the Plague of Cyprian's overall impact on the Roman Empire, we differ rather little from Sabine Huebner's convincing and clearly stated broader assessment (below)—and then only at the analytical margins. Margins, however, that we explore through the mechanism of complex systems theory in Part IV. Huebner states:

Harper argued that the Cyprianic Plague had "significant social, economic, political and cultural ramifications," but this remains a question for debate, since the sources do not easily lend themselves to such a wide-ranging interpretation. While the epidemic undoubtedly exacerbated the political and military crisis of the third quarter of the third century, it should not be considered as the root cause of the crisis itself...

Harper rightly terms it a pandemic though, that spread over the entire Mediterranean world. As is so often the case, this pandemic was probably very disparate in its impact—heavily affecting some densely settled cities... while sparing some more rural areas. It seems that the port city of Alexandria would have been more severely affected than its remote hinterland, and the same was probably true for other major cities [20, p. 170].

We also did not attempt an analysis/modeling, as much out of a choice to keep the project within optimally achievable and practical parameters as from concerns over the immediate feasibility of such an effort, of the plague's potential specific impact on the Roman military—itself an additional highly complex and challenging calculation. We do believe, however, that such an effort would be potentially analytically significant in a broader historical sense related to a better understanding of the Roman Empire's Third-Century Crisis. We may undertake it in the near future. However, it can be inferred that the Plague of Cyprian's impact in terms of overall demography (despite precise calculations being likely impossible), and even if the effects were far more devastating in urban than rural settings, was extensive and particularly ill-timed. The pandemic and the compounding multifaceted Third-Century Crisis that provided its temporal context also probably collectively limited any sort of robust and rapid demographic recovery for the Roman Empire. The demographic, economic, political, and military feedbacks of de-urbanization from plague-related partial or full *ville morte* events must be particularly considered when attempting to generate a picture of the plague's overall effects on the third-century Roman Empire. The damage to such key economic, military, demographic, and administrative nodes, whether temporary or permanent, whether caused by the pandemic alone or by a combination of other factors, must also have produced multiple and mutually amplifying disruptive feedbacks. In this broader vein, we will now step back and examine the holistic image, impact and relevance of the Plague of Cyprian revealed by our assembled puzzle pieces in the concluding Part IV.

References

1 Ajelli, M., Merler, S., 2012 "Transmission Potential and Design of Adequate Control Measures for Marburg Hemorrhagic Fever," *PLoS One*. https://doi.org/10.1371/journal.pone.0050948.

2 Apion, 2003 "A Soldier Writes Home, AD Second Century," in Lewis, Jon, ed., *The Mammoth Book of Eyewitness Rome* (Carroll and Graff): p. 374.

3 Banchich, Thomas, 2015 *The Lost History of Peter the Patrician: An Account of Rome's Imperial Past from the Age of Justinian* (Routledge): pp. 113–114.

4 Benes, Anita, 2018 "The Roman Aqueduct at Brigetio," *Dissertationes Archaeologicae ex. Institutu Archaeologica Universiteta de Rolando Eotvos nominate* 3(6) (Budapest): pp. 419–440.

5 Bente, D.A., Forrester, N.L., 2013, October "Crimean-Congo Hemorrhagic Fever: History, Epidemiology, Pathogenesis, Clinical Syndrome and Genetic Diversity." *Antiviral Research* 100(1). https://doi.org/10.1016/j.antiviral.2013.07.006.

6 Borchert, Matthias, Mulanga, Subue, et al., 2006 "Serosurvey on Household Contacts of Hemorrhagic Fever Patients," *Emerging Infectious Diseases Journal*. https://wwwnc.cdc.gov/eid/article/12/3/05-0622_article. Accessed on 3/18/2022.

7 Borhy, Laszlo and Bartus, David, et al., 2016 "From the Baths of Brigetio to a Roman Baths Visitor Center," https://www.muzeumcafe.hu. Accessed on 12/14/2021.

8 Boris-Breskovic, Bojana, 2014 "A Hoard of Bronze Coins from Viminacium and the Province of Dacia from the Village of Bukovac Near Vranjska Banja," *Numizmaticar*, Broj 32 National Museum of Belgrade, (Belgrade).

9 2018 "Brigetio Thermalspa of Komarom," https://www.komthermal.hu. Accessed on 11/14/2021.

10 Carroll, Serena and Nichol, Stuart, "Ancient Common Ancestry of Crimean-Congo hemorrhagic fever virus," *Molecular Phylogenetics and Evolution* 55(3) June: 1103–1110. https://doi.org/10.1016/j.ympev.2010.01.006. Accessed on 9/22/2021.

11 CDC, 2021 "Travelers Health: Ebola and Marburg," https://wwwnc.cdc.gov. Accessed on 8/19/2022.

12 The Colchester Archaeologist, 2013 "Extraordinary "Bustum" Burial Excavated in Colchester," *The Colchester Archaeological Trust Online*. https://www.thecolchesterarchaeologist.co.uk. Accessed on 8/8/2022.

13 Drca, Slobodan, 2004 "Naissus," in Archaeological Treasure of Nis from the Neolithic to the Middle Ages. Crnoglavac, Vesna ed. (Belgrade University). https://www.academia.edu/1123552/Archaeological_treasue_of_Nis. Accessed on 4/12/2021.
14 Eutropius, *Abridgement of Roman History* Trans. John Selby Watson 1853 (London), Book IX, 9.
15 Fernstrom, Aaron, and Goldblatt, Michael, 2013 "Aerobiology and its Role in the Transmission of Infectious Disease," *Journal of Pathogens*. https://www.ncbi.nlm.nih.gov/pmc/articles/PMC3556854/.
16 Formenty, Pierre, 2017 "Introduction to Crimean Congo Hemorrhagic Fever (CCHF)," *WHO*, pp.1–23.
17 Harper, Kyle, 2015 "Pandemics and Passages to Late antiquity: Rethinking the Plague of Cyprian c. 249-270. *Journal of Roman Archeology* 28: 223-260.
18 Harper, Kyle, 2017 *The Fate of Rome: Climate, Disease & the End of an Empire* (Princeton University Press).
19 *Historia Augusta, LacusCurtius* Trans. David Magie 1932. https://www.penelope.uchicago.edu. First accessed on 4/20/2020.
20 Huebner, Sabine, 2021, June "The 'Plague of Cyprian': A Revised View of the Origin and Spread of a 3rd-c. CE Pandemic." Published online by Cambridge University *Journal of Roman Archaeology* 34(1): 151–174.
21 Juhasz, Laszlo, 2017 "Monetary Circulation in the Municipium of Brigetio: The Excavations of Komarom/Szony-Vasarter, pp. 9–23. *Proceedings of the International Numismatic Symposium: Circulation of Antique Coins in Southeastern Europe* Viminacium, Serbia September 15–17, 2017. Institute of Archaeology, (Belgrade) Korac, Miomic, ed. Accessed on 4/5/2021.
22 *Orbis* Scheidel W. and Meeks E 2012: *ORBIS: The Stanford Geospatial Network Model of the Roman World*. Retrieved 20 July 2022 from http://orbis.stanford.edu. Originally accessed on 6/21/2021.
23 Paulus Orosius, *The Seven Books of History Against the Pagans* Trans. Roy Deferrari.1964 Catholic University of America Press (Washington).
24 Petrov, Vladimir, 2022, January "Krymskaia-Kongo gemorragicheskaia likhoradka," *Vektor*. https://www.vector.nsc.ru. Accessed on 7/30/2022.
25 Scheidel W. and Meeks E, 2012 *ORBIS [new version]: The Stanford Geospatial Network Model of the Roman World*. Retrieved 20 July 2022 from http://orbis.stanford.edu.
26 Serretiello, E et al., 2020, September 3 "The Emerging Tick-Borne Crimean-Congo Hemorrhagic Fever Virus: A Narrative Review." *Travel Med Infect* Dis.; 37:101871. https://doi.org/10.1016/j.tmaid.2020.101871. PMID 32891725. Accessed on 9/29/2021.
27 Shaw, Garry, 2016 "Human Blood, Organs, and a Surprising Virus Detected in Ancient Pottery," *Science*. https://doi.org/10.1126/science.aal0490.

28 Stillwell, Richard and MacDonald, William, et al., 1976 "Brigetio (Szony) Komarom, Hungary, *The Princeton Encyclopedia of Classical Sites*. https://www.perseus.tufts.edu Accessed on 6/11/2021.
29 Towner, Jonathan and Khristova, Marina, et al., 2006, July "Marburgvirus Genomics and Association with a Large Hemorrhagic Fever Outbreak in Angola," *Journal of Virology*. https://doi.org/10.1128/JVI.00069-06. Accessed on 8/9/2021.
30 Vasic, Miloje, 2013 "Contributions to the Study of Denarii from the Period of the Reign of Gordianus III," *Numizmaticar* 32, 2014 National Museum of Belgrade (Belgrade, 2014).
31 Vojvoda, Mirjana and Petrovic, Slavoljub, 2011 "Novac Bitinijske kovnice nikeje iz Banovog Polja u Macvi," *Numizmaticar* Broj, 29, 2011 National Museum of Belgrade (Belgrade, 2011).
32 Zosimus, 2004 *The New History*, Taylor Anderson, ed. (Odin's Library Classics).

PART IV

"Conclusion—Final Thoughts on the Plague of Cyprian": Methodological Defense and Brief Overview of Our "Solution", Historical Context and Current Relevance

"The good thing about Science is that it's true, whether or not you believe in it."—Neil deGrasse Tyson, 2021.

CHAPTER 10

Situating the Plague of Cyprian Within the Broader Outlines of Roman History

Brief Historiographical Positioning, Methodological Defense, and Overview

Table 10.1 summarizes our team's findings about the Plague of Cyprian and compares them to those of the two other prominent scholars who have recently considered the Plague—Dr. Kyle Harper and Dr. Sabine Huebner. All three analyses agree that the Plague of Cyprian was a true empire-wide and significant pandemic event. Our conclusions on origins and temporal parameters largely, if not fully, align with Harper's findings. For the most part, we reject Huebner's alternative theories concerning point of origin, epidemiological directionality, and timeline. Our conclusions in terms of timeline also match the fifteen-year outbreak span attributed to Philostratus. This is a relatively minor point, but it was also one that we didn't want to entirely dismiss. There also seems to be at least some broad scholarly agreement concerning a VHF being a potential pathogenic cause of the plague. Overall, Huebner is the most cautious in her conclusions in this regard, with our work being the most specific and conclusive on this point. Harper is in the middle in terms of his level of certainty. Similar overall congruence is also achieved regarding the Balkans as being a particular plague epicenter. Our analysis, in contrast to Huebner's, however, has the Plague being spread *to* the Balkans as resources and manpower flowed *into* the region *from* an already plague-afflicted empire as it was forced to confront a new military crisis in 251

M. Orsag et al., *Interdisciplinary Insights from the Plague of Cyprian*,
https://doi.org/10.1007/978-3-031-26094-0_10

Table 10.1 Comparison of major findings regarding the Plague of Cyprian in the relevant historiography

Author(s)	*Plague of Cyprian = Pandemic?*	*Geographic Origin*	*Temporal Parameters*	*Likely Cause*	*Balkans a POC Epicentre*	*Overall Impact/Epidemiological Footprint*
Harper	Yes	Non-Roman Africa via Egypt	249–262 CE	Probable VHF**(Ebola)	Yes*	Complex Universal Pestilence
Huebner	Yes	Balkans	251–270 CE	Uncertain (Possible VHF**/Ebola?)	Yes*	More Limited Largely Urban
IHPH/Bucknell	Yes	Non-Roman Africa via Egypt	248–262 CE	Highly Probable VHF-MARV Lineage***	Yes*	More Limited Largely Urban

* For unstated or varying reasons in each case
** Viral Hemorrhagic fever
*** Marburg virus
**** Plague of Cyprian
Table by Dr. Mark Orsag

CE; this is implicitly similar to Harper's relevant conclusions. Contrastingly, however, our more limited "guardrail" modeling of the POC's third-century epidemiological footprint in Part III aligns far better with Huebner's finding of a largely urban-centered pestilence and somewhat more limited appraisal of potential broader impacts as opposed to the more "universal pestilence" and somewhat greater impact appraisal put forward by Kyle Harper (Table 10.1).

We believe that we have produced a probable, multifaceted, detailed, mutually supporting, comprehensive, and holistically interdisciplinary "solution" to the mystery of the over 1,700-year-old Plague of Cyprian. Using the pioneering work of Dr. Kyle Harper (our intellectual and scholarly debt to him is referenced by our quote at the beginning of Part I) as both a guide and a starting point, we have built our case through a multidisciplinary overlay of ancient and modern evidence. Dr. Sabine Huebner's erudite work also proved highly informative and uniquely valuable. On the whole, our interdisciplinary team has attempted a somewhat more consilient approach than either of these two more highly trained classicists (though their research also contains interdisciplinary elements). Our goal was a complete systems synthesis, using a more experimental approach, that incorporated such scientific techniques as modeling and analysis grounded in detailed consideration of molecular-phylogenetic evidence. We have thus widened the methodological approaches that have been traditionally used to examine events such as the Plague of Cyprian. We have also consciously designed each part (as well as the whole) of our work to look at all, regardless of disciplinary boundaries, directly and indirectly available pieces of a shared and unitary puzzle. The overlay of modern interdisciplinary scientific evidence (and modeling based on it) not only fills in missing pieces that the existing fragmentary ancient historical evidence doesn't provide but also creates instances of "experimental" supporting congruence/consilience with it. With surprising consistency, the pieces seemed to fit rather seamlessly and precisely. Yet, there are also many important questions regarding the Plague of Cyprian that our novel systems-synthesis approach simply was not geared toward answering. McNeill, in 1976, referring to Cyprian's *De Mortalite*, found it "intrinsically persuasive" but not "really proven" that, "Such sublime capacity to cope with the horrors and psychic shock of unexampled epidemic was a significant part of the attractiveness of Christian doctrine for the hard-pressed populations of the Roman Empire" [21, p. 109]. This is certainly a not unimportant question that directly concerns the place of the Plague of Cyprian in the broader progression of world history, but we cannot claim to have advanced a more definitive interpretation of it in any way.

Methodological defense—The Challenges of "Thucydides Syndrome"

In 2021, Dr. Kyle Harper wrote a thoughtful essay, "The Two Cultures: Classics and Science in a Time of Pestilence" that noted progress concerning interdisciplinary innovation, and resistance to it, within the field of Classical Studies. Though the field had made great strides in terms of the digital humanities and much research was slowly becoming increasingly interdisciplinary almost by a kind of academic osmosis, entrenched notions of "two cultures" separating the sciences and humanities remain, in Harper's view, as entrenched as ever in some important ways. Such divisions, rooted in the highly specialized nature of modern academic training, differences in epistemological approaches, divisions over the desirability of co-authorship, conventions of academic writing, even differing physical locations on the campuses of institutions of higher learning all effectively "make it hard to collaborate" across disciplinary boundaries. This has led to continuing skepticism within Classical Studies toward interdisciplinary work that bridges gaps with the natural sciences; despite the fact that it is becoming ever clearer "how much the sciences and humanities need each other" [14].

If any event embodies some of the inherent challenges in both effectively collaborating across disciplinary boundaries and of retrospectively diagnosing historical pathogens (particularly in ancient times), it was a rather fruitless decades-long scholarly/scientific debate, and reaction to it, over the ancient Plague of Athens (labeled the "Thucydides Syndrome" by the physicians and scientists who tried to retrospectively diagnose it) that struck in 430 BCE during the Peloponnesian War. It was famously described by the ancient historian Thucydides. His account of the pestilence became essentially canonical in Antiquity, constituting a Greco-Roman cautionary *topos* concerning pestilence and potential resulting social disorder. Broadly, in twenty-first-century historiography, this case can be interpreted in at least two ways. The first, in our view, is highly constructive and analytically enabling as opposed to limiting. It stresses attention to the relevant socio-cultural historical contexts (…and constructs) as well as the consequences of disease. In our discussions throughout of Greco-Roman medicine, for example, with its concepts of humourism, varying schools of medical thought, miasmas, etc., as well as in our discussion of the Plague of Cyprian's broader impacts upon the Roman Empire, we have followed this approach herein.

But if "Thucydides Syndrome" is conversely used as a central argument that retrospective pathogen diagnosis is effectively impossible (short of a definitive DNA test finding), even when attempted by physicians or scientists, we thoroughly reject this argument as both overreaching and limiting, unfounded and distorting (particularly in regard to the repeatedly aforementioned issues regarding RNA pathogens). A central element of this latter argument involves the inconclusive debate over retrospective diagnosis of the pathogen involved in the Plague of Athens. In this single case, a convincing retrospective diagnosis of the pathogen involved indeed proved impossible (even when attempted by successive teams of medical experts). This is despite a highly detailed description of its' pathology in Thucydides' account. The titles from a sampling of the articles in question are highly informative in and of themselves: "The Thucydides syndrome, A new hypothesis for the cause of the plague of Athens," *New England Journal of Medicine*, 1985; "The Thucydides syndrome: another view," *New England Journal of Medicine,* 1986; "The 'Thucydides syndrome reconsidered: new thoughts on the 'Plague of Athens'" *American Journal of Epidemiology,* 1994; "The Thucydides Syndrome: Ebola Deja vu (or Ebola reemergent)" *Emerging Infectious Diseases*,1996; "The good and the bad dying indiscriminately: the Athenian plague reconsidered." *Pharos Alpha Omega Honors Medical Society,* 2000, etc. The debate obviously produced no consensus. A wide range of diseases—including bubonic plague, smallpox, typhus, toxic shock syndrome, (and later) Ebola were all alternatively proposed and strongly challenged [5, 11, 20, 22, 26]. Clearly, historians who expressed skepticism about the efforts of the "trespassing" doctors and scientists to retrospectively diagnose "Thucydides Syndrome," had a very valid point in a narrow sense. While the cultural and scholarly aspects of this case do highlight the need for analytical caution regarding retrospective pathogen identification, should their analytically limiting verdicts be negatively universalized (beyond this undoubtedly baffling individual instance) within the disciplines of History and Classical Studies? With this very sort of methodological challenge in mind, Table 7.1 made the connection, in great detail, between pathology described in the ancient sources *for* the Plague of Cyprian and its striking overlay with the pathology of the Marburg virus.

Table 10.2 tackles the same issue from the opposite analytical direction and constitutes an attempt to demonstrate a "dog that didn't bark in the night" form of consilience. It lists the prominent symptoms for the two earlier proposed disease alternatives for the POC and for the Plague of Athens/pestilential *topos* that *don't overlap with* and *don't match* prominent MARV pathology. It cross-references these with the *corpus* of the ancient sources regarding the Plague of Cyprian. The strongly prevailing *lack* of an overlay, in this instance, was quite telling.

Yet in light of this "Thucydides Syndrome"-based methodological challenge, we wrestled with the question of how to further demonstrate the fundamental validity of our complex methodology, succinctly, via a second, simpler, and easier to analytically digest, example. This was particularly important, we felt, in light of the inherently vast implausibility of any future "definitive" scientific solution to the Plague of

Table 10.2 Comparative differential pathology (POC Ancient Sources): Measles, Smallpox, Plague of Athens not Prominent in MARV [4, 30–32]

	Symptoms	*Noted by Multiple Sources*	*Noted by Singular Sources*	*Not Noted by any Sources*
Smallpox	Full body rash			X
	Pustular scabs			X
	Pock marks			X
Measles	Runny nose			X
	Cough			X
	White spots*			X
	Full body rash			X
Plague of Athens	Sneezing			X
	Loss of voice			X
	Pustules			X
	Ulcers			X
	Extreme Thirst**		X	
	Insomina			X

* Inside cheeks

** Gregory of Nyssa

Table by Dr. Mark Orsag

Cyprian. We also wanted to examine some of the other larger questions at issue. Once again, we rack up an intellectual debt to Dr. Kyle Harper in the process. The central topics in the quotations and samplings below relate to the second-century Antonine Plague (observed and documented by the famed Roman physician Galen), as well as a fourth-century disease outbreak described by Eusebius, referenced earlier in Part I. These outbreaks are both characterized by Harper as likely caused by smallpox. Harper further refined and updated his case on the former based upon the most recent paleo-genomic findings in his 2021 work *Plagues upon the Earth: Disease and the Course of Human History*. First, a passage from the *History Compass* authors, referencing J.F. Gilliam's article "The Plague under Marcus Aurelius," which appeared in the *American Journal of Philology*, back in 1961...

> Yet for many scholars, the Antonine pandemic is nearly back where Gilliam left it in 1961...Gilliam critically assessed the written sources for the Antonine Plague. He concluded, there is not enough evidence to identify the disease responsible, or to trace the epidemic's origin and spread with much accuracy, or to determine even approximately the number of those who died... Gilliam's conclusion still stands...
>
> Harper's selective reading of the scholarship also facilitates his acceptance and use of the smallpox diagnosis. There is no time here to get into the ongoing debate about whether smallpox or a poxvirus ancestor anything like *variola major* existed in late antiquity... [Harper] accepts the results of maximalist work on the Antonine Plague that has been heavily criticized to validate a smallpox diagnosis of the plague. [10, II-pp. 2–6]

We felt that it was useful to include an excerpt from Gilliam's original supporting footnote at this point...

> H. Haeser, *Lehrbach III*, pp. 24–33, concluded that the disease was probably smallpox. H. Zinsser, who discusses the effects of this and other epidemics in antiquity concurs; *Rats, Lice and History* (Boston, 1935), p. 137. A. Castiglioni, *A History of Medicine* (New York, 1941), p. 244 suggests, without conviction, exanthematous typhus or perhaps bubonic plague. Clearly, it is best for **laymen** [emphasis ours] to leave the question entirely open, as does George Sarton, *Galen of Pergamon* (Lawrence, 1954), p. 22... [9, p. 227, (note 8)]

Now, on to excerpts from Kyle Harper's *The Fate of Rome: Climate, Disease, & The End of an Empire* (2017). As Harper notes, Galen also described characteristic windpipe ulceration, and the "black pustules spreading and covering the entire body that then dried and fell off like scales." (Table 10.3).

> A smallpox infection is the closest match to the disease Galen observed. It is worth reviewing in some detail the course of infection by the virus, *variola major* as it was, by modern clinicians working around the globe in the decades leading up to the eradication of the disease...
>
> Galen's observations are consistent with the symptoms of *variola major*. In the course of smallpox infection, death comes around 10 days after symptoms appear, in line with Galen's belief that days 9-12 were critical. The fever was universal but not unusually intense, a credible description of smallpox... The conclusions of a study by Littman and Littman remain sound: Galen had seen the [rarer and far deadlier] hemorrhagic... smallpox. The sign of it for Galen was very black stool, a harbinger of the worst. Galen assumed [from the perspective of the Greco-Roman "bodily humors" medical theory] that the excess of black bile experienced by all plague victims could be expressed as the drying black rash or the passage of bloody stool. The former gave the patient reason to hope, while the latter meant that the blood had been "completely cooked" ... But the rash he described, from first appearances of protrusive lesions to scabbing and scarring, points to smallpox as the agent of the Antonine Plague, even across the cultural chasm that separates us from the ancient doctor...

Table 10.3 Dr. Kyle Harper's adapted table from the *The Fate of Rome*, Detailing "Galen's Pestilential Rash"

Greek	*Description*
Exanthemata melana	Black extrusive/pustular rash...
Helkos	Wound, sore. For Galen, it implies a rupture in the continuity of the flesh...
Ephelkis	Scab. The natural hardening of a sore or wound
Lemma	What is peeled off. The word is used of fish scales. For Galen, the victims whose rash did not turn into sores were scratched off like scales
Epouloo	To scar over. Used frequently by Galen, to cicatrize

[Table courtesy of Dr. Kyle Harper and Princeton University Press [12, p. 104]

> Molecular clock dating is a method of estimating how long ago an evolutionary event occurred: it provides a probability of how long ago it may have taken for a given level of genetic variation to have developed. One placed the divergence of smallpox from its most recent common ancestor with *Tatera poxvirus* in Africa only 2,000-4,000 years ago. [12, pp. 104–107]

Now, the relevant passage from Book IX of Eusebius' *Ecclesiastical History,* describing an early fourth-century pestilence that Harper dates to 312–313 CE and identifies [in our view both absolutely correctly] as smallpox. Harper, moreover, uses this passage quite effectively to differentiate that later disease's pathology from that of the earlier Plague of Cyprian [12, p. 141]. We have included some additional passages from Book IX Chapter 8 of the *Ecclesiastical History* here for reasons that will become clearer below:

> The customary rains and showers of the winter season ceased to fall in their wonted abundance upon the earth and an unexpected famine made its appearance, and in addition to this a pestilence, and another severe disease consisting of an ulcer, which on account of its fiery appearance, was appropriately called a carbuncle. This spreading over the entire body, greatly endangered the lives of those who suffered from it; but as it chiefly attacked the eyes, it deprived multitudes of men and women of their sight...these things suddenly took place at one and the same time...
>
> Some, wasted away like ghosts at the very point of death, stumbled and tottered here and there, and too weak to stand fell down in the middle of the streets: lying stretched out at full length, they begged that a small morsel of food be given to them, and with their last gasp, they cried out "Hunger!"
>
> ...In this way death, waging war with these two weapons, pestilence and famine, destroyed whole families in a short time, so that one could see two or three dead bodies carried out at once. [8, p. 180 (9.8.1-8 and 12)]

And lastly, relevant modern medical information from the Centers for Disease Control on smallpox and smallpox disease types...

> A person with smallpox goes through several stages as the disease progresses. Each stage has its own signs and symptoms...
>
> Early Rash

> ...By the fourth day, the skin sores fill with a thick, opaque fluid and often have a dent in the center. Once the skin sores fill with fluid, the fever may rise again until scabs form over the bumps.
>
> Pustular Rash and Scabs
>
> ...After about 5 days, the pustules begin to form a crust and then scab. By the end of the second week after the rash appears, most of the sores have scabbed over.
>
> The scabs begin to fall off, leaving marks on the skin. Three weeks after the rash appears, most scabs will have fallen off.
>
> There are four main clinical forms of smallpox [Modified-type smallpox is not applicable in our context, so only three apply...], each with different characteristics.
>
> During the smallpox era, the case fatality rate differed for the different clinical forms, but it was approximately 30% overall for unvaccinated individuals.
>
> Ordinary Smallpox (*variola major*)
>
> Ordinary smallpox was the most common form, accounting for over 85% of the cases during the smallpox era.
>
> Eruptive Stage
>
> ...rash lesions begin to develop. Lesions usually first appear on the oropharynx, followed by the face and extremities, and then spread to the trunk, palms and soles...
>
> Sequelae
>
> Pockmarks and scarring are the most common sequelae...
>
> Other sequelae include:
>
> Blindness is a rare occurrence and results from corneal scarring following keratitis or corneal ulcerations. Generally, it occurs when malnutrition and/or opportunistic infections are present. [3, 4]

The conclusions that we can draw from the cumulative overlay of all this material are numerous and quite profound in relation to supporting the validity of our overall methodological approach in a number of ways...

- Gilliam's brief and limited historiographical listing of conclusions regarding the Antonine Plague fails to really engage analytically with the evidence in the way that Harper did. While we will not attempt to comparatively judge our own analysis in this regard, it is notable that two of the main authors of this work hardly, on multiple grounds, qualify as mere "laymen."
- In the words of the *History Compass* authors, Gilliam's conclusion "still stands" regarding the Antonine Plague. But should it be so?

Gilliam wrote at a time when the science of molecular phylogenetics was in its infancy. Findings from that rapidly developing field in the last 60 years seem highly pertinent. These should not be ignored out of concern for artificial disciplinary boundaries or for any other reason. Applicable knowledge is applicable knowledge.

- Implications that molecular phylogenetics is hopelessly imperfect, or irrelevant in this regard is, in our view, very wrong. There are still complexities and lingering elements of "unsettled science" to be sure. But, as we believe that we demonstrated in Part II, these can be analytically handled. In this context, glycoprotein analysis largely replicates WGS. While temporal results obtained from NERVEs, for example, differ from those of WGS, at times dramatically, one has to understand that, in context, the ensuing results regarding events like ancient plagues effectively *reinforce* existing WGS-based linkages to modern diseases by pushing the evolutionary stages of viruses further back in a temporal sense.
- Such views, as expressed by the *History Compass* authors, thus come across as effectively "rejectionist" of the evidence derived from recently emerging but valid science. Such an overly narrow, discipline-abridged view also seems unfortunate, in a practical sense, for the field of history at this moment. This is a broader reality referenced, perceptively we think, by the *History Compass* authors themselves. They note public perceptions of the humanities as being "aloof and disconnected from everyday realities..." [10, II-pp. 2–6 and I-p. 3]. Not only does such a limiting approach forgo a context in which science can clearly inform history, but it also conversely misses an opportunity to have history inform, as in these cases related to smallpox, and provide crucial confirmation/clarification for science!
- The overlay of the modern descriptions of the disease of *variola major* from the CDC and the ancient evidence from Galen and Eusebius align to a truly remarkable degree. Those accounts clearly seem to describe *variola major* or something closely akin to it in an evolutionary lineage sense. The scabbing, rapid spread (implying aerosolization), progression of the disease, the fiery rash, and "carbuncles" are all shared between modern science and the two ancient accounts. There are further modern medical overlaps with the account of Galen—the eruptive pustules, windpipe ulcerations, sores, and (eventually peeling off) scabs... In both ancient cases, the disease

also could be quite deadly, but wasn't necessarily so—there were apparently high percentages of survivors. This is indeed consistent with the CDC's average historical CFR of 30% for the *variola major* form of smallpox. It could also, however, be indicative of a biologically similar but slightly less virulent disease from the smallpox lineage [12, p. 104 and 8, p. 180 (9.8.1)].

- Such ancient–modern evidentiary consilience demonstrates that Harper and WGS were correct, and that smallpox or something very similar from its evolutionary lineage did exist in late antiquity. The same validation now also applies to RNA viruses with the CCHF viral peptide isolations referenced in Part III. One last point in Eusebius' account of the 312–313 CE outbreak is also so perfectly on point in support of Harper's identification of smallpox in that instance that he couldn't have, in our opinion, produced better if he had time-traveled back to the early fourth century and co-written that section of the *Ecclesiastical History* with Eusebius! Blindness is the second-listed *variola major sequela* by the CDC (after only scarring), but is also specifically conditional upon certain accompanying circumstances and described as more generally "rare" otherwise. *Yet, Eusebius' account provides the exact context* (a massive famine causing widespread severe malnutrition) in which that "rare," in the opinion of modern medical science, *sequela* would have been far more abundant. Eusebius, lacking modern medical knowledge, would obviously have had no way to "tailor" that account. For us, this confirming evidence pushes the latter example "beyond a reasonable doubt" especially when twinned with Eusebius' description of the symptomatology. Narrowing this analysis even further, infectious disease more broadly is not a prominent cause of blindness. In the modern developed world, chronic diseases like diabetes (causing macular degeneration and cataracts) are far more prominent in that regard. The bacterium *chlamydia trachomatis* is the leading infectious cause of blindness worldwide today but only accounts for only 1.4% of annual cases of blindness worldwide. It too can be ruled out here. It disproportionately affects children and takes years of gradual (individual) repeated infections to blind. Measles can also cause blindness as a *sequela*, but this phenomenon very disproportionately affects children; something that goes directly against Eusebius' account—"...it deprived men, women and children of their sight." Neither pathology thus bears any resemblance to the rapidly attacking and indiscriminate disease described by Eusebius [12, p. 104 and 8, p. 180 (9.8.1), 33].

- By 2021, paleo-genomic discovery of a (now extinct) smallpox lineage ancestor from the early Middle Ages, which shared an MRCA (labeled "ancient smallpox") with *variola major* around 1,700 years ago, brought Harper's directly verified scientific trail of smallpox lineage pathogenic bread crumbs even closer to the result of both his earlier analytical differential inferences and the time of Eusebius (with the second-century age of Galen, Marcus Aurelius, the Marcomannic Wars, and the Antonine Plague approaching, counterintuitively, in the rear view). Based on the newer scientific data, the shared MRCA (ancient smallpox) seems to be the most likely smallpox lineage pathogen of the Antonine Plague. The smallpox lineage outbreak that was recorded by Eusebius at the beginning of the fourth century could easily have been any of the three—with weighting strongly, in our view, toward the now-extinct milder medieval version (possibly its outbreak debut) due to what seems to have been a lower described CFR despite the very likely highly negative impact of the accompanying famine on the overall immunological wellbeing of the population. Such recent paleo-genomic evidence also allows Harper to convincingly account for two mysterious claims or silences in Galen's and other ancient accounts of the second-century Antonine Plague (animal infections and lack of a clearer description of characteristic post-infection scarring) that differ somewhat from what is known about the more modern version of *variola major*. Thus, it is arguable that the 1,600-year phylogenetic lineage shadow of the Antonine Plague allowed eighteenth-century British physician Edward Jenner (actually building upon existing variolation practices), to be credited with the development of the world's first true vaccine, derived from animal-based cowpox, to inoculate against smallpox in humans. We would also further note that any severity of pathology differences between the Antonine Plague and later smallpox lineage outbreaks could also partially stem from the likelihood that, in the second century, the disease was "new"—having just arrived in the Roman Empire. By the early fourth century, the disease (or at least its evolutionary pathogenic lineage) would have been endemic within the Roman Empire. We could be dealing with a single smallpox lineage pathogen in the two outbreaks, or more than one; it is not altogether certain. Yet, the consilience of the multidisciplinary holistic analysis very strongly implicates the, pathologically and epidemiologically broadly similar, collective evolutionary lineage of smallpox in both cases. *Which has, again, been our overall point regarding the MARV lineage and the Plague of Cyprian exactly.* Despite a certain general similarity in the nature of the three ancient disease mysteries, the better and more "focused" direct scientific evidence relating to smallpox means that the cases can be made far more succinctly in these (second and fourth century) ancient outbreak puzzles than for our more complex third-century case. The holistic and painstaking multidirectional consilience that we built regarding the temporally intervening Plague of Cyprian pandemic is largely unnecessary in these bookending instances [15, pp. 194–196].

- The overall result of the ancient-modern overlay here, in our view, based on these second and early fourth-century examples, not only *fully* validates both Kyle Harper and molecular phylogenetics (and the closely related field of paleo-genomics), but also strongly supports the general veracity of the ancient accounts of Galen (despite the distorting influence of prevailing erroneous Greco-Roman medical theories) and Eusebius, whose likely hyperbole in portraying the suffering of the pagans through famine and pestilence as divine vengeance for the persecution of Christians didn't apparently distort his description of the disease itself. Both ancient commentators were seemingly quite accurate. Harper's achievement of an even more complete-historical scientific consilience in *Plagues,* supplemented by our own analysis here, also pushes the dial for the implication of the smallpox lineage in both instances past a "beyond a reasonable doubt" standard in our view. The early fourth-century outbreak, in particular, indeed becomes an analytically air tight, if still not scientifically confirmed, case.
- Our larger, somewhat more complex case regarding the Plague of Cyprian has many of the same overall strengths, such as being mutually supporting and achieving a sort of time-bending (Harper in *Plagues* aptly uses the phrase "time-travel") consilience. It also shares one, in context, excusable, weakness—an unavoidable tendency to be somewhat circular or reductionist in a limited and narrow sense. We have mitigated this intrinsic but limited problem with our overlay and modeling methodologies that repeatedly provided

further mutually reinforcing and multidirectional interdisciplinary consilience.

As our work progressed, we began to conceive of our results as measured against five ascending levels of proof. Level 1 (basic plausibility) and level 2 (comparatively likely in contrast to any existing other explanations); these we believe that we have clearly attained. Level 3 would be a probable but far from certain "preponderance of the evidence" standard. That we believe that we have also solidly obtained. At the higher end of the scale, however, we freely acknowledge that our circumstantial case, no matter how plausible or even probable, does *not* approach a level 5 standard of scientific proof, or quite rise to a level 4 "beyond a reasonable doubt" evidentiary standard. For us to be broadly wrong, however, a long string of unlikely coincidences would have to be true. With such a consistent, holistic, and multidirectional "fit" of our pieces, such a scenario seems, to us, unlikely. Yet, parts of our case are quite speculative, contingent upon others and there also remain some lingering, if not necessarily significant, doubts and uncertainties. More broadly, possible reductionism and circularity unavoidably lurk as the analytical downsides of multidisciplinary and consilient approaches such as ours. Ongoing scholarly debate over the Dionysius/Eusebius account in Part I (though we believe our argument there is strong...) and the partial conflation (we admittedly couldn't solve it...) between modeled outbreak timing in Aksum and pathogen dissemination to the Roman Empire in Part II are two specific uncertainties/problems. We could add the inability to *entirely* rule out Ebola or pandemic influenza to this list. More broadly, a variety of uncertainties and gaps bedevil the ancient evidence, particularly in Part II. We also lack a single clear irrefutable "smoking gun" in terms of science or history. Even some areas where we seemingly have abundant evidence, however, can be, at least slightly, questioned. For instance, what if Cyprian (who lacked modern medical training obviously) got some key disease symptoms wrong or conveyed mistaken impressions through use of humourist concepts, religious metaphor/allegory, or ancient *topoi* in his accounts? Here, we once again bump up against the sort of concerns raised by "Thucydides Syndrome."

On the other hand, however, this latter seems a less than credible scenario in that Cyprian's "distortions," whether linguistic, perceptual, or simple errors of fact, would "coincidentally" end up aligning so strongly (in terms of both specific points [positively, negatively, and holistically]) with the pathology of a filovirus as deduced by modern medical science. It seems comparatively far more likely that he, like Galen and Eusebius, was largely reporting accurately the details of the pandemic horror that he saw around him. And yet, the cumulative effects of destruction and chaos in third-century antiquity (and for centuries and millennia after), climate, the inherent fragility of RNA viruses (leading to a near complete lack of preserved material from them); the analytical challenges inherent in interpreting ancient descriptions of disease through a veil of erroneous, complex (and imperfectly understood) humourism, imprecision in temporal placement of ancient events and simply the broader and multifaceted effects of the passage of millennia complicate matters. As noted, it took the somewhat unlikely circumstance of bodies frozen in the Alaskan tundra to definitively solve the modern, barely century-old pathogenic mystery of the Spanish flu pandemic. "The sheer unlikelihood that ancient biomolecules are preserved" is thus an ongoing and inescapable problem. The "gold standard" of scientific proof, in the case of the Plague of Cyprian, is very probably (remarkable paleo-genomic finds involving viral peptide proteins isolated from ancient pottery shards aside) unreachable in the foreseeable future [15, p. 47].

Still, a hopefully consensus-shifting and "comparatively likely" answer that expands our knowledge and challenges our analytical skills in terms of making interdisciplinary sense of pandemics (historical or otherwise) is far better than none at all. As we have assembled our multidisciplinary portrait of the Plague of Cyprian from this overlay of the ancient and modern evidence, the overall relative analytical straightforwardness of the effort seems evidence in and of itself. The existing ancient and modern pieces almost all fit seamlessly. Though we found no evidence that served to directly undermine our proposed solution, many other pieces of the puzzle are still missing. Our interconnected result felt incomplete and somewhat inferential, yet highly supportable in a multidimensional and multidisciplinary evidentiary sense, complex yet naturally fitting. Barring some sort of near-miraculous archeological or other evidentiary discovery, however, we believe that our case has nearly reached its promising, and for us highly probable, but still frustratingly elusive "end of the road." And, with the exception of a more detailed systems-synthesis examination focusing on the applicable epidemiology and pathology-management of the Roman military environment, there seems little realistic hope for our reaching anywhere beyond our current limits.

Despite this project's undoubted limitations, we believe that our work still makes significant scholarly and analytical contributions in several senses: (1). It contributes (building upon the scholarship of Kyle Harper, Sabine Huebner, and others) to a more holistic and contextualized understanding of the Plague of Cyprian, a significant ancient pandemic event. (2). The work's novel interdisciplinary methodology offers a new approach to examining and analyzing phenomena, such as historical pandemics, that we hope will prove useful to future multidisciplinary teams of scholars and be further refined and perfected by them. (3). This book contributes to a greater awareness (and also to awareness of the limits of such understanding) of the place of the Plague of Cyprian within the broader context of the history of the Roman Empire. (4). Our work also broaches potentially important ideas concerning the interconnectedness of human disease and ecology (an embodiment of the vast and unified complex concept of "planetary health") as well as those of complex systems theory concerning civilizational collapse. In this latter vein, we believe that the work offers larger interdisciplinary insights that are relevant and applicable to not just the third century but also to the twenty-first. It is to these two latter points (#s 3 and 4) that we now turn...

Historical Holistic Context

As Kyle Harper has noted in his analysis of the historical role of pandemics, "Throughout history, epidemic events have always been conjunctions of chance and structure." They can "coincide with other crises– climate, legitimacy, monetary, armed conflict– they mark moments of transformation or redirect the stream of history" while at other times pandemics "simply accelerate or reveal where we were already going." More specifically, the Plague of Cyprian, in the third century, was part of a multifaceted constitutional and monetary crisis that left the Roman state profoundly altered and the geopolitical posture of the Romans vis-a-vis the Persians and the Germans systematically weakened. But in a sense, these changes were visible before the pestilence, which hastened them along" [13]. While a more detailed analysis of Rome's Third-Century Crisis lies beyond the scope of our present work, we (once again) strongly concur with the more limiting latter part of Harper's analysis. The beginning of the Third-Century Crisis, in our view, pre-dated the pandemic.

The overthrow and assassination of the generally effective and apparently well-intentioned Emperor Alexander Severus by Maximinus Thrax in 235 CE, and the civil war that followed a few years later, highlighted the growing potential for political instability, civil conflict, and military anarchy. These recurring phenomena would prove some of the key hallmarks of the next several decades. Similarly, the military shortcomings that the entire context of Alexander's assassination revealed also anticipated the uncharacteristic military failures that would beset the Roman Empire during the early and middle stages of the Third-Century Crisis. The empire's brewing concurrent monetary and financial crisis, which Alexander's effective policies seem to have temporarily ameliorated, also seems to have worsened as a result of his murder.

That a crisis atmosphere prevailed is evidenced not only by the ancient literary sources but also on the basis of the supporting overlay of modern-derived patterns of prevalence in terms of numismatic finds. Comparisons of the Croatian Kamenica and Deleu hoards, for example, produced a recent scholarly conclusion that they dead ended at and were thus "probably hidden during… [Maximinus Thrax's] reign." This is also true of a number of other regional numismatic finds in the Balkans, such as those of Ravna, Supska II, Postojna, Sanadinovo, and Csapon. Another find from the village of Mehovine in Serbia consisted of 2,000 silver coins that had apparently been secreted by the owner of a third century *villa rustica*. The probable reason again lies in Maximinus' confiscatory fiscal policies and the "violence" of his soldiers and agents in carrying them out. Romans indeed seem to have been desperate to avoid the extortionary practices of their gigantic (reputedly 6 ft. 8 inches tall) ruler whose taxes and confiscations, in the words of the *Historia Augusta*, "reduced the richest men to utter poverty" and whose officials "despoiled everyone ruthlessly." Thrax's costly indulgence of the soldiers, to whom he owed his power, and his tactically effective but strategically futile and costly wars in Germany not only drained the imperial coffers but also threatened the prosperity of Roman citizens—the emperor desperately needed revenue and "dispatched agents throughout the whole empire to collect it." The Third-Century Crisis' phase of "military anarchy" had thus already arrived, at a very fundamental and socially broad level, more than a decade before there even was a Plague of Cyprian [1, 17, 23, ("The Lives of the Two Maximini, (13.5–14.1)].

Yet, the Roman Empire ultimately survived all of these challenges, including the de-urbanizing semi-apocalyptic Plague of Cyprian. Key to recovery from the multiple impacts of the Third-Century Crisis, in an organizational and military sense, was the brave and highly capable soldier-Emperor Aurelian (r. 270–275 CE), who was appropriately nick-named *Manu ad Ferrum* (Hand on Hilt). The *Orbis Restitutor* (Restorer of the World), as he was also known, rose from a surprisingly humble and provincial background through the ranks of the understrength and struggling legions. Even in the assessment of Orosius, a hostile Christian source, the pagan Emperor Aurelian was "a man of extraordinary ability in the affairs of war." Aurelian indeed displayed seemingly superhuman military and administrative skill and courage in driving back successive waves of invaders and politically re-uniting a splintered empire. This was something that, before him, had seemed utterly impossible. Aurelian, for many Romans, seemed the living embodiment not of Apollo the Healer but rather of Sol Invictus—the unconquerable warrior sun god under whose worship ("one faith, one empire") the emperor sought to culturally re-unite the Empire. In rebuilding the empire's trading, transport, food supply, and reserve networks, fortifying the city of Rome and unsuccessfully (his one failure) battling runaway inflation, Aurelian expanded the military and effectively militarized whole sectors of the economy. While Aurelian's successes marked the end of the worst of the Third-Century Crisis, the Emperor's assassination, instigated by the duplicity of the corrupt courtier Eros in 275 CE, showed that the multifaceted crisis that the Plague of Cyprian had greatly exacerbated had not yet been entirely weathered [27, p. 319 (7.23), 17, ("Life of Aurelian")].

Aurelian's reign, however, also marked the beginning of a broader and more multifaceted imperial transformation. The Principate's last remnants crumbled as the Third-Century Crisis eased. Rome had survived plague, civil strife, invasions, and political fragmentation. In response to these challenges, a new rigid and massive top down bureaucratic-military structure arose to maintain order and cope with new challenges. Aurelian's successors, such as the emperor Carus, "did not even ask Senators to acclaim" them. Emperor Diocletian's (r. 284–305 CE) later more transformative and comprehensive reforms, however stabilizing, also codified and accentuated these trends [16]. Diocletian was followed by an all-powerful and "transformative" tyrant—Constantine the Great (r. 306–337 CE). He reversed course in cultural-religious matters by seeking unity in what had been a source of division and dissent; he began the Christianization of the Roman Empire. Constantine also geographically re-centered it; he created an impressive Eastern capital at Constantinople (in modern-day Turkey) and instituted a new gold currency, the *solidus*. The Roman Empire became structurally finely balanced, somewhat more ruralized (probably partially as a result of the Plague of Cyprian), and even more chronically financially overstretched. Though, in the fourth century's "Age of Gold,'' a fair measure of the pre-Third-Century Crisis prosperity, somewhat paradoxically, returned. The surviving empire's "sheer fiscal power remained historically exceptional– on par with the most formidable of [much later and more technologically advanced] seventeenth-century polities." Diocletian's intervening not fully successful social mobility, taxation, and wage and price controls may well, in light of our findings, have been a product not only of a desire to control inflation but also to control geographic mobility and flight to the disproportionally Plague of Cyprian-depopulated cities whose (necessary) demographic rebound nevertheless endangered the critical labor demands of agriculture if uncontrolled. Such a scenario may essentially have anticipated the structurally similar Late Medieval/Early Modern European migratory phenomenon that later followed the Black Death [18, 24, pp. 294–298].

Social mobility, however, in the rigid system indeed became far more limited and the late Roman elite (in yet another precursor of the Middle Ages to come) became essentially hereditary. The peasantry became tied to the land and its aristocratic owners. The *speculatores*, (army scouts) as well as those employed by the Roman postal service, were transformed under Diocletian into a vast network of secret police, spies, and informers. Much of the civilization associated with the traditional religious, cultural, and socio-political values of the city of Rome itself (which declined in status and population) had been profoundly transformed. The direct and indirect impacts of the challenges of the Third-Century Crisis– pandemic, invasion, political and military anarchy, etc., and Rome's adaptive responses to that combination of existential threats, thus collectively feedbacked, evolved, and rippled on through time:

By the mid-fourth century, the burden of taxation became ever more unequal and crushing for the common citizen as a top-heavy bloated bureaucracy and a massive, but less effective, military establishment drained the empire's, still considerable, financial resources. The wealthy sheltered behind tax immunities, and anti-inflationary policies pursued by a series of emperors only partially succeeded. Emperors, such as Diocletian, were known as *Dominus Noster* (Lord and Master). Subjects, in rituals of submission borrowed from the undiluted autocratic traditions of Rome's Sassanid Persian enemies, had to prostrate themselves in the emperor's presence (*adoratio*) with the most favored getting the opportunity to crawl forward and kiss the hem of his robe. Many Romans saw this as a humiliating, alienating, and "empty spectacle" [28].

In this traditional and harsh scholarly as well as oft-popularized interpretation, Rome's semi-democratic and constitutionalist traditions, derived from the Republic and even from the *Principate*, were now seen as a threat themselves—to a precious and demonstrably fragile order and stability that events such as the terrifying Plague of Cyprian, chronic civil wars, and Germanic and Persian invasions had nearly shattered. A radical political transformation had occurred as a reaction to the plague, invasions, climate unpredictability, civil wars, inflation, and other threats of the Third-Century Crisis and beyond. Manpower shortages also continued to bedevil the late empire—in particular its restructured and increasingly "barbarianized" military. The growing apparatus of the increasingly state-affiliated but still badly divided and quarrelsome Christian Church was effectively also a major competitor for the military. Traditional Roman patriotism eroded; the exploited lower classes most acutely seemed to have no recognizable stake in or benefit from the imperial system and no incentive to defend it [19, pp. 363–367]. Even during more prosperous periods, such as the reign of Constantine the Great, wealth tended to flow mostly to the elite—with the rest of the population left behind. Attempts by successive emperors to ameliorate the economic disparities and to, culturally or religiously, unify the Roman Empire through a single faith proved only partially successful. Christianity proved to be a very durable faith (it would survive the empire's fall and dominate the Middle Ages), but fourth–fifth-century Christians were too theologically divided to establish a truly single orthodoxy and become an ideal force for Roman unity. Pagan and Jewish resistance to Christianization also sporadically continued throughout the fourth century, causing considerable conflict [7, p. 237]. In addition, the influential landed nobility was very reluctant to part with their *coloni* laborers. With military conscription drives failing as a result of all these factors, successive later emperors turned to cheaper barbarian mercenaries to fill the ranks of the legions and additionally reorganized Rome's legions with an eye chiefly to containing spiraling defense budgets. The qualitative edge that had been traditionally enjoyed by the Roman Empire in the military sphere thus began to erode by the middle-to-late fourth century. To quote Lewis again:

> The later Roman Empire, in other words, was a bankrupt militarized state with an ineffective army, a parasitic inverted aristocracy, and a hostile peasantry. Large parts of it ceased to be "Roman" at all, as barbarian tribes settled under official imperial approval and then with no approval at all. [19, p. 366]

Such sentiments, while broadly correct in certain ways, are also now characterized by more recent scholarship as overly simplistic. Scholarly consensus regarding the late Roman Empire has become somewhat less Gibbonian and condemnatory and rather more nuanced. Dr. Jan Willem Drijvers of the University of Groningen perceptively summed up the current overall scholarly verdict on the late Roman Empire—that which emerged transformed from the Third-Century Crisis—in the following way:

> Frontiers have really shifted. Classicists, ancient historians and archaeologists have become aware that late antiquity is a distinctive period that stands on its own. The later Roman Empire is considered a vital and strong society, characterized by both continuity and transformation. Our knowledge about late antiquity has increased considerably through a flood of publications...Although there has recently been criticism of the paradigm of continuity and transformation, because it insufficiently takes into consideration aspects of crisis, decline and collapse.... [6, p. 2]

Such an alternative, and now paradigmatic, "transformationalist" perspective is most closely associated with the heralded work of Dr. Peter Brown (such as *The World of Late Antiquity* [1971] or *Through the Eye of a Needle* [2012]), but has been adopted in full or in part by many other classicists. Yet, like the old more wholly negative assessment, the more recent scholarly consensus, with its nuanced and less negative assessment of the late empire and analysis concerning Rome's Fall (or Falls), has its critics and challengers. Scholars have approached the "Fall" of the Roman Empire with differing perspectives, but no unitary understanding of one of the most complex and impactful collapses in human history seems likely to emerge. In her recent work, *The Falls of Rome: Crises, Resilience, and Resurgence in Late Antiquity*, Dr. Michelle Salzman challenges the notion of both a temporally unitary "Fall" and "catastrophist" (neo-Gibbonian) notions of uninterrupted "decline and fall" in relation to the collapse of the Western Roman Empire. Linking the fortunes of the city of Rome and traditional Roman values to the fortunes of the

Italian Senatorial aristocracy as opposed to the office of emperor, Salzman highlights the endurance and resilience of Roman aristocrats in the face of a series of challenges. These included the 410 CE sack of Rome by Alaric, the difficult aftermath of the devastating 455 CE occupation of Rome by the Vandals, the Civil War of 470 CE (which she actually sees as a far more devastating event than the traditional Fall—the deposition of Emperor Romulus Augustulus by Odoacer in 476 CE), the highly destructive twenty-year Ostrogothic-Byzantine wars of the sixth century, and finally the fragmenting effects of the Lombard invasions after the Byzantine reconquest of much of Italy. In the end, Salzman identifies the poorly documented dissolution of the Senate in the early seventh century as the final, albeit cumulative, "Fall of Rome." This event seems to have implicitly resulted from increasing Byzantine hostility, particularly from the Eastern Roman Emperor Phocas (r. 602–610 CE), Lombard military pressure and steadily increasing Papal power and influence within the city of Rome itself [2, 29]. As Salzman eloquently concludes:

> When the Eighth-Century *Donation of Constantine* granted the Roman clergy the privilege of wearing senatorial insignia such as the black slippers and white stockings, they had no competition from living senators. Rather, these were simply fitting reminders of the senatorial power that had once existed in this city now firmly under papal control. [29, p. 336]

Salzman's analysis is indeed superbly sourced and analyzed, as well as being very convincing from its chosen perspective, in terms of analyzing Rome's "Falls." Salzman's viewpoint, even though emphasizing "change over rupture" is also not purely "transformationalist." She acknowledges "harsh breaks created by certain crises." If Salzman is a "nuanced transformationalist," in context, our perspective constitutes a mirror image of sorts to hers—"nuanced catastrophism." Her work is nonetheless an effective cautionary tale for "catastrophists" of all varieties, and yet... Salzman's analysis seems to, indirectly at least, de-emphasize the collapse of the Western Roman Empire outside of Italy and some of the complex destructive feedbacks from those events for the city of Rome itself. She is aware of and does not dismiss either; they just aren't the focus of her analysis. [29, pp. 13–21]. Just as Salzman's perspective offers a balancing and cautionary note to Harper's; Harper's "catastrophism" offers an equally eloquent cautionary note to Salzman's "neo-transformationalism":

The history of the fifth century would see the progressive fracturing of the [Roman military] power and the piecemeal loss of hegemony in the west." Former provinces like Britain simply fade from view, thrown on their own local resources, while others like Africa, were hijacked in the full light of day." Some settlements– the Goths in Aquitaine, the Burgundians in Savoy and the Ostrogoths in Italy– were administered with a certain degree of legal nicety. But the empire was dealing from a position of despair… Everywhere but narrow corridors of Italy and Gaul, the machinery of power in the west ceased to be Roman. [12, p. 195]

And the consequences of these wider changes created indirect but significant feedback of their own in and around Rome itself. Modern isotope-based scientific research conducted at the archaeological site of ancient Portus, "the maritime port of Imperial Rome" revealed that after the Vandals' seizure of Roman Africa in the 430s and their Sack of Rome in 455 CE, the diets of Portus' dockworkers, which had been quite diverse and featured many imported food products, became far more subsistence level and less varied as the Western Roman Empire collapsed—resulting in a "breakdown of Roman control of the Mediterranean." Though, in turn, a "transformationalist" might credibly argue that the diets of the dockworkers showed resilience and adaptation to the new circumstances and merely became more similar to those of the rural coloni/peasants [The Middle Ages were beginning to evolve out of late antiquity…] surrounding them. As always, the Western Roman Empire's collapse seems complex enough to defy a simple or unitary analytical interpretation [25]. The Fall or Falls of Rome appear to be, in certain ways, too large and complex of an event to be holistically comprehended and yet still analyzed in detail—even with the benefit of hindsight and from a far more advanced twenty-first-century "analytical height." Might we, however, nevertheless extract relevant modern lessons overall? Or learn from analyzing more specific events, like the Plague of Cyprian, that might have been contributing factors along the way within this larger holistic collapse progression?

References

1 Boris-Breskovic, Bojana and Vojvoda, Mirjana, 2012 "Hoard of Roman Silver Coins from the Village of Mehovine near Sabac," *Numizmaticar*, Broj 30 National Museum of Belgrade.

2 Brown, Peter, 2012 *Through the Eye of a Needle: Wealth, the Fall of Rome, and the Making of Christianity in the West, 350–550 AD.* (Princeton University Press).
3 CDC 2016 "Smallpox: Clinical Disease," *Centers for Disease Control and Prevention* https://www.cdc.gov. Accessed on 5/16/2021.
4 CDC 2017 "Smallpox Virus: Signs and Symptoms," Centers for Disease Control and Prevention https://www.cdc.gov.
5 David, E.M, 2000 "'The Good and the Bad Dying Indiscriminately': The Athenian plague reconsidered," *Pharos Alpha Omega Alpha Honor Medical Society* PMID: 10918791.
6 Drijvers, Jan Willem, 2011 "The Culture of Imperial Leadership in Late antiquity," *Pennsylvania State University* p. 2 *academia.edu.*
7 Ehrman, Bart. 2018 *The Triumph of Christianity: How a Forbidden Religion Swept the World.* (Simon and Schuster), p. 237.
8 Eusebius Pamphilus, *Ecclesiastical History* (Trans. A.C. McGiffert), 1890 Pantianos Classics.
9 Gilliam, J.F, 1961 "The Plague under Marcus Aurelius," *American Journal of Philology*, 94: 225–251, p. 227, fn. 8.
10 Haldon, J, Elton, H, Huebner, S, et al., 2018 "Plagues, Climate Change, and the End of an Empire: A Response to Kyle Harper's *The Fate of Rome, History Compass/Wiley* Parts 1–3[inclusive]. https://doi.org/10.1111/hic3.12508. https://doi.org/10.1111/hic3.12506. https://doi.org/10.1111/hic3.12507.
11 Halladay, A.J, 1986 "The Thucydides Syndrome: Another View," *New England Journal of Medicine* PMID: 3531865.
12 Harper, Kyle, 2017 *The Fate of Rome: Climate, Disease & the End of an Empire* (Princeton University Press).
13 Harper, Kyle, 2020 "The Coronavirus is Accelerating History to the Breaking Point," *Foreign Policy.* https://foreignpolicy.com/author/kyle-harper. Accessed on 10/24/2020.
14 Harper, Kyle, 2021 "The Two Cultures: Classics and Science in a Time of Pestilence," September https://www.classicalstudies.org. Accessed on 8/5/2022.
15 Harper, Kyle, 2021 *Plagues upon the Earth: Disease and the Course of Human History* (Princeton University Press).
16 Hekster, Oliver (with Zair, Nicholas), 2008 *Rome and its Empire, 193–284.* https://www.academia.edu/958257/Rome_and_its_Empire_AD-193-284?email_work_card=view-paper.
17 *Historia Augusta, LacusCurtius* Trans. David Magie 1932. https://www.penelope.uchicago.edu. First Accessed on 4/20/2020.
18 Kagan, Donald, Ozment, Stephen, et al., 2007 The Western Heritage: Volume B: 1300–1815 9th edition (Pearson), pp. 294–298.

19 Lewis, Jon, ed. 2003 *The Mammoth Book of Eyewitness Rome* (Carroll and Graff), pp. 363–367.

20 Longmuir, A.D, 1985 "The Thucydides Syndrome, a new Hypothesis for the Cause of the Plague of Athens," *New England Journal of Medicine.* https://doi.org/10.1056/NEJM198510173131618.

21 McNeill, William H, 1976 *Plagues and Peoples* (Anchor Books).

22 Morens, D.M, et al. 1984 "Thucydides Syndrome Reconsidered: New Thoughts on the Plague of Athens," *American Journal of Epidemiology* 1984 PMID: 10918791.

23 Nad, Miroslav, 2017 "A Part of the Kamenica Hoard from the Sipus Collection," *Proceedings of the International Numismatic Symposium: Circulation of Antique Coins in Southeastern Europe.* (Viminacium, Serbia) Institute of Archeology, Belgrade, Korac, Miomir, Ed. pp. 49–72, p. 58.

24 Nelson, Lynn Harry, 2018 "The Reforms of Diocletian, 284–305 CE," *Brewminate.com* October 2018. https://brewminate.com. Accessed on 7/31/2022.

25 O'Connell, Tamsin, Ballantyne, et al., 2019 "Living and Dying at the *Portus Romae* June 2019," *Antiquity* 93(369). https://doi.org/10.15184/aqy.2019.64.

26 Olson, P.E, et al., 1996 "The Thucydides Syndrome: Ebola Deja vu (or Ebola reemergent)," *Emerging Infectious Diseases* 2, pp. 155–156 ISSN-1080–6059.

27 Paulus Orosius, *The Seven Books of History Against the Pagans* Trans. Roy Deferrari. 1964 (Catholic University of America Press, Washington).

28 *Rome: Power and Glory*, 2010 Parts I-VI. https://www.imdb.com.

29 Salzman, Michelle, 2021 *The Falls of Rome: Crises, Resilience and Resurgence in Late Antiquity* (Cambridge University Press).

30 Thucydides, *The History of the Peloponnesian War* 2.48.1. https://www.gutenberg.org.

31 WHO, 2021 "Marburg Virus Disease,". https://www.who.int. Accessed on 7/26/2022.

32 WHO, 2019 "Measles: Key Facts/Overview,". https://www.who.int. Accessed on 7/25/2022.

33 WHO, 2022. "Trachoma,". http://www.who.int. Accessed on 7/25/2022.

CHAPTER 11

Modern Relevance of the Plague of Cyprian

While the multifaceted debate over Rome's complex Fall or Falls between "transformationalists" and "catastrophists" will certainly continue, there is also an ongoing interconnected debate about the overall cause(s) of civilizational collapse. This is true especially as it pertains to the Western Roman Empire as a leading example of that phenomenon. In such a discussion, it is imperative to both provide working definitions of civilizational collapse as well as to outline and then apply the framework of systems dynamics with the recognition that societies are, themselves, nothing more than complex, adaptive, social-ecological systems (SESs), which are based on the concept that humans are a part of, not separate from, nature. When defining such civilizational collapse, we will first defer to eminent collapse theorist Dr. Joseph Tainter of Utah State University, who states that, "A society has collapsed when it displays a rapid, significant loss of an established level of sociopolitical complexity." This can be characterized by less social stratification, less economic and occupational specialization, population decline, less centralized control and de-urbanization, governance units with smaller territories, less sharing and trade, and less investment in the elements that often define "civilization" such as achievements in art, literature, and architecture. While this definition of simplifying collapse excludes the environment, it is not completely separate from ecological collapse in which an "ecosystem suffers a drastic, possibly permanent, reduction in carrying capacity [maximum population

M. Orsag et al., *Interdisciplinary Insights from the Plague of Cyprian*, https://doi.org/10.1007/978-3-031-26094-0_11

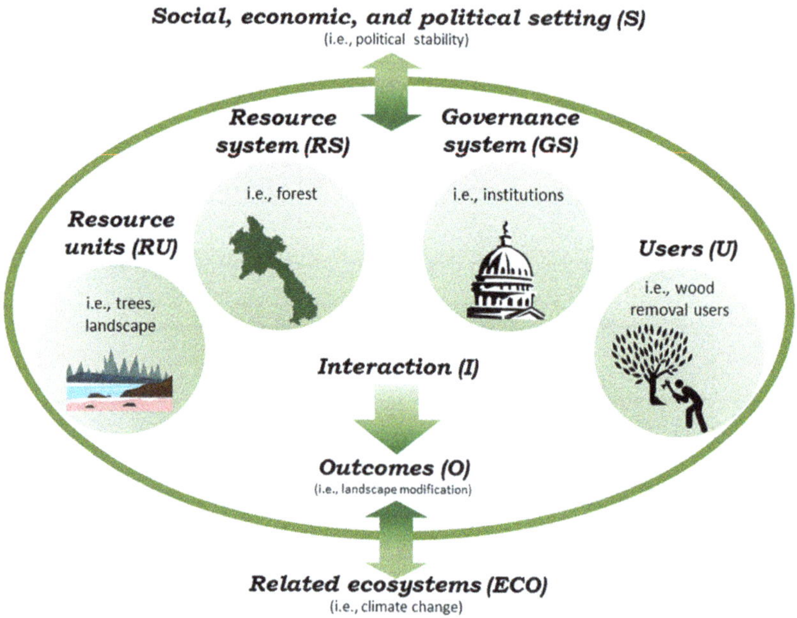

Image 11.1 The social-ecological system framework [public domain]

size of a biological species that can be sustained by that specific environment given the food, habitat, water and other resources available] for all organisms, often resulting in mass extinction." This is due to the fact that the population decline seen in civilizational collapse is essentially due to a reduced carrying capacity of a degraded ecosystem [29] (Image 11.1).

> The social-ecological system framework as described in Ostrom [2009]: "Socio-ecological systems reflect a highly interconnected relationship between society and ecosystems. Resilience [the ability to continue to function in the face of perturbations] of such a system depends on a wide range of factors stemming from the linkages between human societies and ecosystems. The factors include changes in the social, political and environmental situations. These factors cause stress and disturbance, which a community may overcome if it is truly resilient. The interaction between actors of both subsystems further complicates the matter and increases the vulnerability of the system. The concept of socio-ecological resilience embodies

> the capacity of linked social–ecological systems to absorb recurrent disturbances in order to retain essential structures, processes and relationships" [5].

Researchers in the discipline of complex systems dynamics have revealed that the adaptive cycle (below) can be applied and utilized to understand the self-organizing nature and evolution of complex adaptive systems, including SESs, that occurs over time. At times, in a true systems-synthesis manner, they have applied models drawn from engineering related to the physical collapse of complex structures. In the words of Ugo Bardi, civilizational collapse is, as in engineering, a "feature, not a bug," indicating that all civilizations (i.e., complex, adaptive SESs) follow the adaptive cycle in which the system passes through phases of exploitation (or growth), conservation, release (or collapse), followed by reorganization which opens the door to a new phase of growth. The most common example of the adaptive cycle in nature is that of an old growth forest in a state of conservation undergoes collapse or release by way of fire. The destruction is succeeded by new growth of faster, more opportunistic plant species and eventually by trees that, once again, grow into a mature forest and enter a new conservation phase. Applying the adaptive cycle to society has the potential to be extremely useful for gaining perspective on political and social processes through the recognition that these processes are components of a complex **system** that is constantly changing and reorganizing, sometimes quite significantly [3, p. 1; 28] (Image 11.2).

> Social-ecological system dynamics can be represented by an adaptive cycle with four distinct phases: growth or exploitation, conservation, collapse or release, and reorganization. The adaptive cycle exhibits two major phases or transitions. The first, from exploitation to conservation, "is the slow, incremental phase of growth and accumulation." The second, from release to reorganization, "is the rapid phase of reorganization leading to renewal" [Resilience Alliance].

The application of the adaptive cycle to the Roman Empire is not novel, however, it may be more innovative and useful to apply adaptive cycles to smaller systems within civilizations and empires and view them as a panarchy, or a set of nested adaptive cycles interacting with one another (Image 11.3).

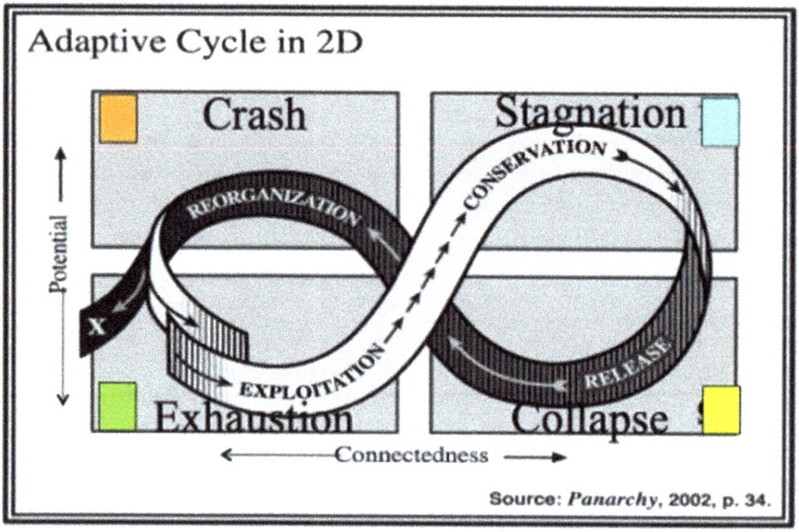

Image 11.2 Panarchy: *Understanding Transformations in Human and Natural Systems*, 2002 (Gunderson, L.H and Holling, C.S (eds.) [Public domain])

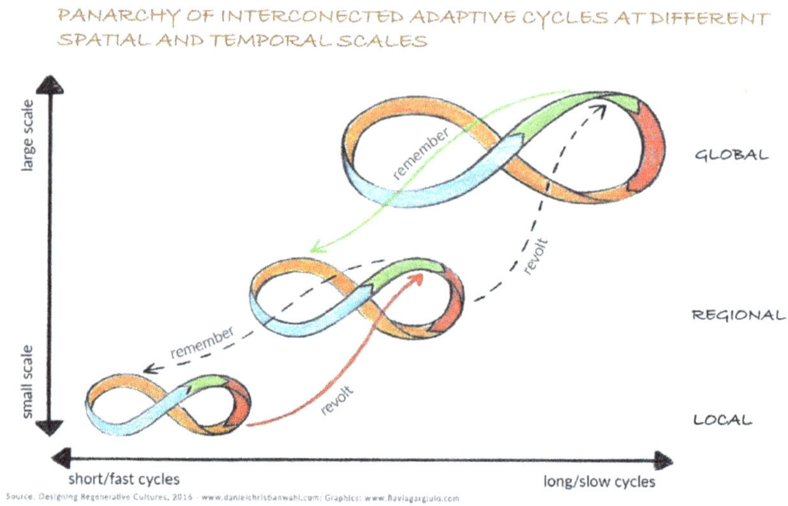

Image 11.3 Panarchy of Interconnected Adaptive Cycles [Public domain]

This model allows for a more global or world-systems analysis considering the "whole and over long periods of time... A world-system is not necessarily planetary in scope, but does necessarily include multiple societies or polities, with long-term, highly structured *interrelationships.*" Systems dynamicists Chase-Dunn and Hall, when applying the panarchical model to society write that, "The most important unit of analysis for the study of social change is not societies or states but the entire world-system. Changes in organization are not endogenous to individual societies. Rather, they are a consequence of complex interactions among local, regional, societal and global processes." This model allows for the ability to view events, like the Plague of Cyprian, as a smaller but more rapidly moving, adaptive cycle that is generally adding momentum to the larger, more slowly moving cycle of the empire- a civilization-based adaptive cycle. It is also important to recognize that this model may also be, perhaps, an incomplete explanation of "why" civilizations/social ecological systems inevitably follow the adaptive cycle and collapse. In the end, we would argue that civilizational collapse often occurs partially as a response to ever-increasing levels of complexity and diminishing returns upon that complexity. But not all collapses are similar; collapse of complex civilizations is, unsurprisingly, a complicated and varied historical phenomenon. It can be validly argued that Rome's Fall or Falls perfectly embody such divergence. The Western Roman Empire yields an example of ultimately uncontrolled "hard" collapse, but the troubled but still wealthy Eastern Roman Empire, contrastingly, shrunk, simplified, transformed, adapted, and struggled but ultimately showed greater resilience—surviving for nearly another millennium. It thus demonstrates an example of controlled "soft" collapse. Though fragmentary and greatly diminished, but in some ways still complex and sophisticated, the Eastern Roman Empire's fortunes waxed, waned, and gradually evolved/devolved into what became known in the Western world as the Byzantine Empire [4].

In our view, the works of systems theorists such as Joseph Tainter and Ugo Bardi are directly applicable to consideration of these questions. In the second chapter of his fascinating and thought-provoking work *The Seneca Effect*, Bardi presents a compelling meld of his views and Tainter's as they relate to "The Mother of All Collapses: The Fall of Rome":

Tainter notes that complexity is a necessary factor for a large social structure to maintain itself. Clearly, he refers to all the structures that are dedicated to command and control of the system: the police, the army, the

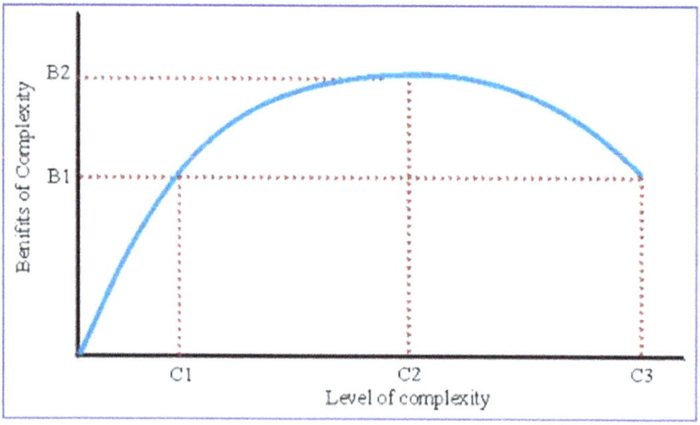

Graph 11.1 Dr. Joseph Tainter's Concept of Diminishing Returns of Complexity (Image courtesy of Dr. Ugo Bardi and Springer International 2017 [3, p. 17])

judiciary system, the bureaucracy, the imperial court, and more... Tainter notes that complexity has a cost that increases as the system becomes larger and needs a larger and more complex control structure. The crucial point of Tainter's idea is that complexity shows diminishing returns to scale, a well-known characteristic of economic systems. That is, when the control system becomes very large, its returns don't increase in proportion. On the contrary, they diminish. [In the figure, you see how Bardi describes Tainter's concept in graphical form] (Graph 11.1).

But how would the phenomenon of diminishing returns of complexity be related to the decline and the collapse of civilizations and empires? According to Tainter, it is because society tends to solve the problems it faces by always increasing the size of its control structures [3, pp.16–17].

To further expand on the concept of diminishing returns of complexity and how it relates to collapse, the example of the Roman army can be considered. The late Roman army (or armies of the Western and Eastern Empires) was/were enormous, complex, and geared toward specific types of imperial border defense. And yet, when put to the test between 376 and 378 CE in an unexpected way during what was originally more a Goth refugee crisis than an actual invasion, the highly complicated Roman military and civilian bureaucracies repeatedly failed to cope with the seemingly minor threat of a force of Goth warriors who likely numbered

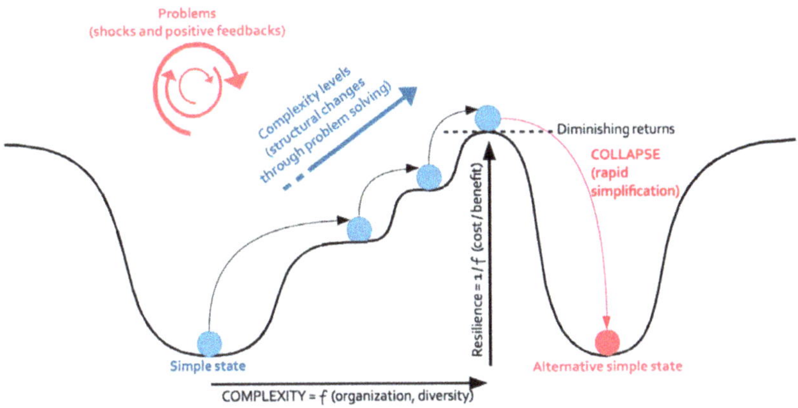

Image 11.4 Rising Civilizational Complexity to "Hard Collapse" (Rapid Uncontrolled Simplification) (*Source* Evans, Henri-Count. (2017). Public Understanding and Perceptions of Climate Change and Global Warming in Zimbabwe and South Africa [Public Domain])

less than 20,000 men. Local government corruption, differing military roles between the mobile field armies and stationery *limitanei*, different emperors in East and West who were colleagues yet rivals, seemingly differing priorities and even conflicting viewpoints of imperial and local governments toward the barbarians and changes in military tactics all combined to lessen the efficiency of Roman efforts. In the words of historian Alessandro Barbero, "It may seem strange that an empire that maintained an army of 500,000 or 600,000 men could not gather enough troops to crush an army of rebels that could not have [initially] numbered more than twelve thousand…" This theme of military over-complexity and ineffectiveness has been echoed by other prominent historians such as Adrian Goldsworthy. Complex systems precisely designed to carry out the intricate and finely balanced task of cost-effectively defending the vast empire in all its geographical "immensity" thus played a key role in leading and contributing to its undoing at the Battle of Adrianople in 378 CE [2, pp. 48–103 (71) and 6]. Similar examples of this same basic cycle of increasing complexity and diminishing returns on it also abound in the modern world (Image 11.4).

Convergence of complexity and resilience theories for modeling social collapse. In the events of multiple and amplified shocks, system resilience erodes as a result of diminishing marginal return on the cost of complexity.

Yet, that is not the whole of the puzzle. Tainter's theory of "diminishing returns of complexity," however, offers a mainly qualitative explanation for the collapse of complex societies and, to a degree, neglects the recognition of the energy and resources that all complex systems require to function. Expanding our understanding of this, John Michael Greer, in his work "How Civilizations Fall: A Theory of Catabolic Collapse," offers a more quantitative model based on human ecology (how humans relate to their natural, social, and built environments). Such collapse is the result of "a self-reinforcing cycle of decline driven by interactions among resources, capital, production, and waste [includes pollution]." In this model, resources (natural material resources, soil fertility, human and information resources) and capital (physical capital, human capital [such as laborers], social capital, and information capital), via the process of production, create new capital *and* waste. This represents the growth or exploitation phase of an adaptive cycle and, once it has begun, will continue in a self-reinforcing manner unless some limits to growth are present. According to Greer, two factors that can limit growth are resource depletion and the energetic and material costs of maintaining the newly produced capital. Once a society faces the end of growth, two options remain. One option is to work toward a "steady state" economy in which the use of capital stocks remains low enough to also meet the maintenance costs of production while also using resources at or below the rate at which they can be replenished, thus avoiding depletion and excessive maintenance costs as well as helping to prevent ecological collapse from exorbitant pollution. Greer, citing the ecological economist, Herman Daly, defines the steady-state economy as "an economy with constant stocks of people and artifacts, maintained at some desired, sufficient levels by low rates of maintenance throughput, that is, by the lowest feasible flows of matter and energy from the first stage of production to the last stage of consumption..." The aim of which is a stable or very mildly fluctuating level of both population numbers as well as consumption of energy and materials; an even more ambitious version of this type of thinking is advocated by modern "degrowth" theorists [8].

The second option is to attempt to prolong the cycle of growth through acceleration of resource and energy (in modern times primarily in the form of fossil fuels) acquisition through military conquest and/or new technological developments. While this drives production of capital and economic growth, it tends to be counterproductive as it also leads to more waste (as a by-product of capital production) and growth of the maintenance cost of capital as well as exacerbating resource depletion. Once a critical resource is depleted and capital can no longer be maintained or pollution from the production process reaches a critical threshold, growth is no longer possible, and the society enters a state of contraction or even, potentially, catabolic collapse in which society must cannibalize its own infrastructure leading to a simplification of societal activities and both decline and ruralization of the population. Indeed, Greer theorizes that the Western Roman Empire fell victim to a complex interrelated (with other factors discussed previously) version of this latter phenomenon.

The collapse of the Western Roman Empire was a catabolic collapse driven by a combined maintenance and resource crisis. While the ancient Mediterranean world, like imperial China, was primarily dependent on readily replenished resources, the empire itself was the product of an anabolic [growth] cycle fueled by easily depleted resources and driven by Roman military superiority. Beginning in the third-century BCE, Roman expansion transformed the capital of other societies into resources for Rome as country after country was conquered and stripped of movable wealth. Each new conquest increased the Roman resource base and helped pay for further conquests. After the first century CE though, further expansion failed to pay its own costs. All remaining peoples within the reach of Rome were either barbarian tribes with little wealth, such as the Germans, or rival empires capable of defending themselves, such as the Parthians... Without income from new conquests, the maintenance costs of empire proved unsustainable, and a catabolic cycle followed rapidly. The first major breakdown in the imperial system came in 166 CE, and further crises followed until the Western Roman Empire ceased to exist in 476–480 CE [6].

Evidence supporting the, at least partial, applicability of this theory can be found, for example, in the decree of Emperors Leo and Majorian (issued from Ravenna) addressed to Emilian Prefect of the City of Rome in 458 CE. Clearly, the city of Rome's own inhabitants were looting, cannibalistically repurposing, and simplifying/restructuring it. "...Nobody shall destroy or damage the whole set of buildings, that is to say the temples and monuments founded by our forefathers..." Punishments for violations ranged from fines to beatings, to amputating an offender's hands" [18]. The Eastern Roman Empire, however, escaped this type of fate. According to Greer...

The Eastern Roman Empire's territory decreased further with the Muslim conquests of the seventh and eighth centuries CE; while this was involuntary, the effects were the same. Successfully shifting to a level of organization that could be supported sustainably by trade and agriculture within a more manageable territory, the eastern empire survived for nearly a millennium longer than its western twin [6].

By uniting Tainter's theory on diminishing returns of complexity, Greer's model of catabolic collapse, the adaptive cycle, and panarchy, we can start to paint a picture of how pandemics, like the Plague of Cyprian, have the potential to play a dual role in civilizational collapse. They serve as external perturbations with the capacity to push the system toward a collapse-related tipping point—as triggers of the system's ever-increasing complexity and the often costly, repressive, and desperate search for the means to sustain that complexity. In the Roman Forum today, one can gaze upon its largest building, the hulking and crumbling ruin of the Basilica of Maxentius. It is a fitting monument, in a way, to the bureaucratization and complexity of late imperial governance. It is noticeably, however, less intact and visually impressive than some of the smaller yet more striking buildings from the earlier centuries of a healthier empire. Yet, paradoxically, Rome could still build a structure as massive as the Basilica in the fourth century (and felt compelled too...) (Image 11.5).

A striking embodiment of the diminishing returns of complexity of the late Roman imperial super-state that one can still see today are the remains of the Basilica of Maxentius in Rome's Forum. This only partially surviving, yet still massive structure, must have housed innumerable imperial bureaucrats during its fourth–fifth-century pinnacle.

While Rome remained plagued by largely seasonally driven diseases (some particularly severe outbreaks of malaria in particular), massively destructive pandemic events along the lines of the second-century Antonine Plague or the third-century Plague of Cyprian were also notable by their relative absence in the fourth and fifth centuries. However, facing a second "perfect storm" of external perturbations and feedbacks that began in 376 CE with a Goth refugee crisis, the brittle, top-heavy, unmanageably complex, financially stretched, and finely balanced

Image 11.5 The Basilica of Maxentius (The Roman Forum) [public domain]

western half of the divided empire steadily; under ceaseless mutually reinforcing waves of barbarian invasions and internal military, economic, and political fragmentation and growing chaos, collapsed. Whether this collapse had occurred by 476, 480, or 486 CE, or later in the sixth or seventh centuries being, as we have seen, eminently debatable. In the wake of the Western Roman Empire's political demise, the population of the city of Rome probably eventually dipped, from its one-time height of around 900,000 to a million people, to well below 100,000 [14, pp. 190–191; 18 , 13, p. 161; 24].

Pandemics act as one type of external perturbation that can amplify a civilization's existing problems and hasten a society toward a tipping point and collapse, accelerating movement through its adaptive cycle by unleashing a host of cascading negative health, economic, social, and political assaults. Adverse health outcomes include a loss of population and debilitation of some portion of the surviving population via lingering effects of the disease. While not even approximately calculable, in our view, the mortality from the multiple waves of the Plague of Cyprian appears to have been very high in many urban areas. Additional urban depopulation stemming from the Plague of Cyprian was indirect—due to fear and flight from the cities (*ville morte* events). Rapid urban depopulation has severe economic implications including reduced tax revenues, overall logistical capacity, labor shortages, and other fear-induced behaviors resulting in a slowdown of economic activity as well as indirect disruptive impacts upon activities required to meet the basic needs of the populace like agriculture. Other significant social and political impacts such as increased social tensions, discrimination, and political unrest are also common. The ability for a society to be resilient in the face of a pandemic depends on the pre-pandemic level of social, political, and environmental stability. Often, however, pandemics typically only strike when the health of society, its environment, and its people are already compromised. Pandemics usually are human-induced phenomena and the results of a system's expansion, ever-increasing complexity and insatiable appetite for growth. For evidence of this, we can look back to Greer's model. In order to continue growth (population, economic, territory size) of a society, ever-increasing quantities of energy and resources, as well as consumption of (often new or exotic) luxuries into which the generated surplus wealth of elites is poured, are required to run economies (produce and distribute goods) and feed a growing population. Human activities to obtain these energetic and material resources such as deforestation, wildlife habitat degradation and fragmentation, intensification and expansion of agriculture, trade in animal and plant species, and climate change also drive biodiversity loss and enable novel and emerging infectious diseases via increased human exposure to natural carriers of pathogenic microbes through disruption of the balance of ecosystems. As mentioned previously, human health is ultimately dependent upon the health of the planetary ecosystems from which we provision the necessities of life and the spoils of civilizational success. An entire "solutions-oriented, transdisciplinary field and social movement," Planetary Health, has been developed and is devoted to "analyzing and addressing the impacts of human disruptions to Earth's natural systems on human health and all life on Earth" [21]. In our analysis, we theorize that the Marburg pathogen may well have come from human exposure to Egyptian rousette bats carrying the virus inside Aksumite mines. The virus then spread to the Roman Empire due to the intense commercial contacts and voracious Roman demand for resources and exotic, luxury commodities from both Aksum and farther afield (India and China). If we are correct, this constitutes a classic, if pre-modern, example of a disruption between the human–animal–environment interfaces that occur in the face of anthropogenic activities.

Ecosystem disruption and anthropogenic climate change have been accelerated in our modern era, primarily since the beginning of the industrial revolution and use of fossil fuels. However, climate change likely also played an indirect role in unleashing the Plague of Cyprian. The RCO, which lasted from ca. 200 BC–150 CE, was a "phase of warm, wet and stable climate across much of the Mediterranean heartland of empire… [and] an inviting moment to make an agrarian empire out of a pyramid of political and economic bargains. Alongside trade and technology, the climate regime was a silent, co-operative force in the seemingly virtuous circle of empire and prosperity." When the RCO ended and the empire entered the "Roman Transitional Period" (ca. 150 CE), a cooler and drier climatic environment, we see the first major pandemic, the Antonine Plague, in 165 CE. Certainly, the negative impact of this change in the climate would have been felt in terms of agricultural outcomes, having the knock-on effects of a decline in population health via malnutrition in addition to the disruption to the existing ecosystems and biodiversity within. Harper, in his above analysis, notes that Romans were shorter than pre-Roman Italians, and showed a greater incidence of defects of the tooth enamel which are likely due to "malnutrition or infectious disease–or the synergies between them." This characterization may be incomplete and difficult to precisely situate temporally, however, and may also minimize the importance of the diet on immunity. High-fiber, whole plant foods that are also high in antioxidants and other polyphenols are critical for the maintenance of a beneficial intestinal microbial milieu and proper human immune function. Harper also does make the excellent point that "there was no such thing as a 'Roman diet'—only the aggregation of socially and regionally variable diets…". However, as the Roman empire expanded, urbanized, and the specialization of occupation occurred, it would have had a significant impact on the diets of urbanites, in particular. From current scientific literature, and having significant consequences for modern society in terms of the burden of chronic disease, it is now known that urbanization leads to higher intakes of the macronutrients, fat, and protein, with a reduction in carbohydrates that are complex in nature such as whole grains, legumes, and other whole plant foods, compromising the health of the population via increased rates of chronic disease and improper functioning of the immune system. Certainly, this has played a role in the lethality of the current Covid-19 pandemic. Whether or not this process similarly influenced Roman disease susceptibility is, however, far less certain [13, pp. 14–91; 15].

Another oft-inferred potential contributor to Rome's Fall or Falls was resource depletion—of precious metals, agriculturally productive land in the Western Empire, or timber. While modern-day climate change is the result of anthropogenic greenhouse gas emissions from burning fossil hydrocarbons, the end of the RCO was primarily the "work of nature." However, in order to fuel the Roman economy, feed the population, and grow the empire, massive deforestation was undertaken and not without consequence. Deforestation disrupts wildlife habitat, increasing human exposure to novel microbes and disrupts ecosystems significantly enough to influence the regional climate by suppressing rainfall. "The case could be made that Roman deforestation interacted with natural patterns of late Holocene climate change to tip the circum-Mediterranean climate toward a regime with less summer precipitation." While, according to Tainter, deforestation was not a major factor in the Western Roman Empire's Fall, it can be argued that deforestation acted as an amplifier to the intense challenges and the complexity that developed in response. The challenges included nearly continuous wars and rebellions, invasion, disease, an excessive degree of urbanization, population decline, and extension of territory beyond cost-effective ability to defend and administer. In the 2011 environmental book *Life Without Oil* by Steve Hallett, the author argues that the collapse of the Roman Empire may have been linked to a peak wood scenario in the Mediterranean Basin. He suggests that, as wood had to be hauled from ever further away, the law of diminishing returns undermined the economic performance of Roman industries such as mining, leaving Rome, indirectly and financially, vulnerable to the other, well-documented problems of invasion and internal division. Hallett discusses this as a cautionary tale and compares it to contemporary society's potential fate under a post-peak oil scenario [13, p. 52; 29, 12].

In a modern context, the degree to which humanity has shaped the planet to advance our own global civilization is, in some ways, seemingly impossible for the human mind to grasp fully. Recent research has revealed that only twenty-three percent of the planet's land surface (excluding Antarctica) and thirteen percent of the ocean can still be classified as wilderness. As such and unsurprisingly, sixty percent of infectious diseases in humans are now zoonotic. SARS-CoV-2, the virus causing the most recent Covid-19 pandemic, serves as a modern example of this phenomenon, having probably made the leap to humans in a wild animal market in Wu-Han, China. However, other recent examples also include SARS-CoV-1 (SARS), H1N1 (swine flu), Ebola, and HIV (the virus that causes AIDS) as well as the avian flu virus that we now know caused the Influenza Pandemic of 1918 ("Spanish Flu"). What's more concerning is the acceleration of pathogen identification as we continue to encroach on wild habitats. Seventy-five percent of the more than thirty such pathogens that have been detected in the last three decades originated in animals [7, 1].

In fact, in a piece by scientists Morand and Walther, in which they advocate for a voluntary reduction in travel (to reduce the spread of emerging infectious diseases) and economic degrowth (to achieve the steady-state economy discussed previously), the pair write, "The greatly accelerated economic growth during the Anthropocene [the period characterized by socio-economic and earth-system related indicators experienced as a result of exponential growth after the Second World War] has resulted in astonishing improvements in many aspects of human well-being but has also caused the acceleration of risks, such as the interlinked biodiversity and climate crises. Here, we report on another risk: the accelerated infectious disease risk associated with the number and geographic spread of human infectious disease outbreaks. Furthermore, the spatial distribution of these outbreaks is becoming more globalized in the sense that the overall modularity of the disease networks across the globe has decreased, meaning disease outbreaks have become increasingly pandemic in their nature" [21, 16]. While, on a modern scale, the second-century Antonine Plague, the third-century Plague of Cyprian, or the early twentieth century's deadly global Spanish flu would have been full-fledged Category 5 pandemics based on the CFRs (whether directly measured or inferred) of their respective pathogens (pathology strongly informs the modern pandemic scale), the twenty-first century's Covid-19 pandemic is probably a somewhat less apocalyptic Category 2 or 3. Yet, nonetheless, it is a great example of the direct and indirect disruptive/hastening feedbacks for complex societies from pandemics; such outcomes are currently obvious in our daily lives. Covid-19 thus constitutes a great example of why the destabilizing effects of pandemics on human society and careful consideration of their proximate causes, namely human endeavors (including empire building), should never be minimized (Image 11.6).

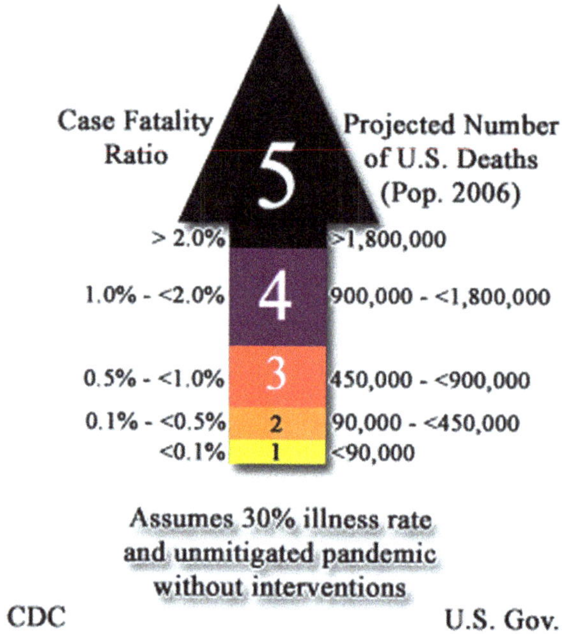

Image 11.6 Adapted from the Centers for Disease Control's Pandemic Severity Index [public domain]

Feeding back with the ever more intricate complexities of our own economic, political, and social structures; with the interrelated vulnerabilities of globalization, climate change, and "just in time" supply chains; with labor demands and socio-economic inequalities; with racial, ethnic and national divisions, with regional tensions and great power rivalries, and most directly with chronic "diseases of civilization" caused by unhealthy diets, this current "mild pandemic" has killed (at the time this last section of our team's work was finalized in late 2022) around 6.7 million people and destabilized our modern globalized world—possibly for decades to come. In the words of medical author Dr. Michael Greger, the current Covid-19 pandemic may also just be a "dress rehearsal" for a possibly far deadlier Category Five epidemic influenza outbreak. Or to use a broader apt quote from complex systems dynamicist, Dr. Nate Hagens of the University of Minnesota, "The future is just the history that we cannot yet see." Zoonotic, highly infectious flu strains with pandemic potential and projected human fatality rates of 50% (H5N1 is a current example) or more are emerging from confined animal feeding operations (CAFOs), particularly for poultry, in China and other nations around the world. One can only imagine the nightmarish effects of such a dramatic "potential ramping up of [pandemic] virulence" on the highly interconnected and complicated globalized supply chains, the food, water, power, and health care systems worldwide that have already been so severely strained by Covid-19. Such a hypothetical yet realistically possible "perfect storm" event could have unparalleled consequences for our species and planet. There are also, of course, the continuing unresolved, increasing, and proven threats from such deadly zoonotic viruses as SARS-CoV-2, Ebola, and Marburg. These vast, and clearly now all too real, zoonotic threats are perhaps best collectively summarized and analyzed in their stunning clarity and complexity in David Quammen's remarkably prophetic book *Spillover: Animal Infections and the Next Human Pandemic* [9, 11, 25]. And yet the threat, in such a scenario, might not only come from the pandemic itself, but also (indirectly) from our potential responses to it.

If the hulking ruin of the Basilica of Maxentius constitutes the physical manifestation of responses to challenges and the embodiment of diminishing returns of complexity, the Plague of Cyprian was one contributor, in the long and complex sequence of third to fifth-century challenges, to the collapse of the Western Roman Empire. The interaction of these events over time was both cumulative and synergistic resulting in an amplification of their effects. The Plague of Cyprian could be considered, in climate science lingo, a "threat multiplier" or an "accelerant of instability," pushing the empire faster along its adaptive cycle. In the former Roman imperial lands throughout much of Western Europe and North Africa, simplicity replaced complexity, as de-urbanization, which began with the Plague of Cyprian during the Third-Century Crisis, was effectively completed. There is certainly evidence, that at least in certain ways, the collapse and simplification involved were "hard" (even catastrophic) in nature. An intriguing cultural-literary rendering of this vast collapse process occurs in such medieval sources as the physically damaged manuscript commonly referred to as "The Ruin." Written in Old English sometime during the Eighth Century, the *Codex Exoniensis* (fols. 123b–124b) laments the collapse, through the description of the destruction/decline of a once-marvelous Roman imperial fortress city (it is unclear which one exactly), of the earlier Romano-British world.

The wondrous city was the embodiment of a society that was the northwestern frontier of a great, intricately interconnected, sophisticated, and almost incomprehensibly "far flung kingdom." At times dismissed by medievalist scholars as a mere elegy, the *Codex's* anonymous author actually seems to long, consciously and directly, for the restoration of the Roman world with its lost unity, prosperity, knowledge, architecture, technology, monetary, cultural and socio-political unity, arts, clever craftsmanship, and engineering. The *Codex* is thus more properly characterized as "an encomium." The faded hydro-mechanical engineering wonders of the Roman baths embody a civilization with superior levels of "livability and comfort" that had long fallen into oblivion in Britain by the eighth century—through a combination of violent destruction and "from age eaten away." Despite its sad decay, the ruined Roman fortress city still evokes, from a medieval perspective, "awe and wonder" and seems to be the work of "giants." This source additionally seems to provide, according to the literary analysis of Dr. Liam Purdon, evidence that at least the later waves (under Valerian and Gallienus) of the Plague of Cyprian impacted British towns and cities producing destructive "days of pestilence" in the middle of the third century. Though as Purdon points out, the *Codex* author's apparently confused, and rather "imaginative," conflation of dim and distant third, fourth, and fifth-century events somewhat complicates the issue. The interconnected, lost, broader, more complex, and greater world, however, had vanished and since descended into a kind of, in the memorable terminology of Nate Hagens, "Great Simplification" [10, 24].

Not only do we, in the Westernized and globalized twenty-first-century world, have the grave threats of lurking zoonotic pathogens and the destabilizing pandemics they spawn in common with the Romans, but we also share a parallel to their "built in," climate-cycle-related, annual rise in late summer/early fall urban mortality. In Rome's case, this was due to malaria, tuberculosis and other infectious diseases, bacteriological food poisoning, parasites, etc. In third-century Rome, these constituted a separate phenomenon to the pandemic and unique Plague of Cyprian. Unlike the ancient Plague of Cyprian, however, the modern *impact of Covid-19 has been devastatingly amplified by its interface* with our "built in," financially draining, and seemingly permanent epidemics of diet-linked chronic diseases of modern civilization such as diabetes, obesity, hypertension, heart disease, and cancer: "The co-morbidities so often cited as heightening the risk of severe infection or death from SARS CoV-2—such as obesity, diabetes and heart disease are inextricably linked to unhealthy diets." While Roman society's urban institutions, such as the baths, the medical system, and the shift toward more elaborate funeral rites likely created, in our view, a "destructive feedback phenomena" with the Plague of Cyprian's MARV evolutionary lineage pathogen in terms of *epidemiology*, such parallel destructive feedback phenomena have been similarly applicable in terms of amplifying Covid-19's *pathology*. The overall structural parallels, once again, fit tragically well. Our twenty-first-century globalized world's high-tech but patchy and uneven medical system has countered Covid-19 in ways that St. Cyprian or even Galen could not have comprehended. Methods such as the rapid creation of multiple effective (if necessarily imperfect against a fast-evolving RNA virus like SARS-CoV-2) and life-saving vaccines have been crucial. Yet, a simultaneously evolving "quiet pandemic," involving 1.2 million "directly-related deaths" in 2019 alone from what had been considered common but "treatable infections" (with another estimated 5 million deaths indirectly linked that year) is due to deadly amplifying feedback *from* the modern medical system. It is as significant a factor in this instance as the Roman medical system was in our modeling of the probable urban functioning of the third-century Plague of Cyprian. The causes of this vast modern annual mortality are a variety of strains of antibiotic-resistant bacteria that evolved in response to over/irresponsible use of modern antibiotics, primarily in the livestock industry. As noted by the University of Washington-led team of researchers that published their findings in *The Lancet*, the annual human toll, centered in but not exclusive to poorer countries, "caused by bacteria resistant to antibiotics" now dwarfs global annual mortality from the relatively "new" threat of AIDS or the age-old scourge of malaria. This has led to scientists proclaiming the arrival of the "post-antibiotic era" when "routine surgery or chemotherapy is considered too dangerous, because there are no drugs to prevent or treat bacterial infections… [potentially leading to] the decimation of modern medicine [13, pp. 72–91; 30, 27, 17]."

Similarly functioning, but possibly far worse, destructive amplifications may loom in the near future. Climate and environmental disruption will also likely act as threat multipliers for those "carrying the burden of noncommunicable diseases" (NCDs) as medical supply chains may become even more severely disrupted in the near twenty-first-century future. Chillingly, 60% of American adults now have "at least one chronic disease, while 40% have two or more chronic conditions." In a twenty-first-century echo of our findings for the Roman third century, excessive urbanization also poses clear risks for increased population susceptibility stemming from multiple causes including "urban heat island effects," which exacerbate chronic disease-related morbidity and mortality. For these enduring problems of the oft-overlooked dangers of the social burden of chronic disease, as well as the interrelated and immediately destabilizing perils of pandemic disease, effective potential responses exist for us in ways that they didn't for the Romans. Science-based dietary guidelines that would "marry our human health goals with climate realities," address other interconnected problems such as industrial agriculture's contributions to the Earth's rapidly deteriorating planetary health and assist a controlled move away from fossil fuels exist, but we have yet to fully employ them [22].

Pathogens indeed have played and still play a complex and crucial role in human history (and vice versa). To close, pandemics like the Plague of Cyprian and Covid-19, striking at whatever level of severity, thus have the capacity to help shape events in three distinct ways: Directly through their demographic impacts; indirectly through their broader complex social, economic, political, etc. impacts; and through the effects of the civilizational responses that they trigger. Likely, these smaller pandemic adaptive cycles merely "push" the larger civilizational cycles along the path that they were already traversing (we believe this to be broadly true of both the Plague of Cyprian and Covid-19); though a countervailing outcome of varying strengths is also, of course possible. Kyle Harper argues for just this oppositional adaptive cycle role for the Eastern Roman Empire with the sixth-century Plague of Justinian. As Harper notes in *Plagues upon the Earth*," Human history shapes disease ecology and pathogen evolution; disease ecology and pathogen evolution in turn [help] shape the course of human history" [13, p. 206; 14, p. 5]. *Ouroboros*...the snake swallows its own tail. We ignore the timeless lessons of this deadly adaptive cycle of pathology, epidemiology, ecology, and history at our peril.

Our holistic reconstruction of the devastating Plague of Cyprian, within the broader "crisis" of the third-century Roman Empire, thus stands as a key component of a multifaceted and highly relevant cautionary tale for the Covid-19 pandemic-destabilized, globalized twenty-first-century world. Like the Western Roman Empire, we now face our own crisis, or rather a "meta crisis," defined by the reality that fossil fuel use is absolutely necessary to maintain our complex, global civilization in its current state but continuing to do so is leading to the potential for civilization-ending climate change and ecological collapse.

As with Rome's Third-Century Crisis and the Plague of Cyprian and Covid-19 in the twenty-first century, pandemic events often display an exacerbating function as subsidiary, interacting, smaller and faster-moving adaptive cycle drivers, pushing threatened civilizations toward faster, and more uncontrolled "hard" collapses (as the Western Roman Empire ultimately did). This, coupled with the current reality that fossil hydrocarbons are finite resources, are in decline, and without them, our economies, levels of consumption, and prevailing civilizational archetype cannot be maintained and will, in some manner, collapse. This has resulted in an all-encompassing human predicament for which there are no solutions, only responses. Our responses, as a species, will have to be multidisciplinary, complex, and bold to help find the resilience to soften (as the Eastern Roman/Byzantine Empire ultimately did) our own civilization's landing as we travel along our own continuing adaptive cycle as a species.

References

1 Allan, James, Venter, Oscar, et al. 2017 "Temporally Inter-Comparable Maps of Terrestrial Wilderness and the "Last of the Wild," *Scientific Data*, https://doi.org/10.1038/Sdata.2017.187
2 Barbero, Alessandro, 2005 *The Day of the Barbarians: The Battle that led to the Fall of the Roman Empire* Trans. John Walker (Walker, New York), pp. 48–103 [71].
3 Bardi, Ugo, 2018 *The Seneca Effect* (Cham, Switzerland: Springer), p. 1.
4 Chase-Dunn, C and Hall, T.D. 1997 *Rise and Demise: Complex World Systems* (Westview).
5 Francis, Royce, and Bekera, Beheila, 2014 "A Metric and Framework for Resilience Analysis of Engineered and Infrastructure Systems," *Reliability Engineering and Safety*,https://doi.org/10.1016/j.ress.2013.07.004.
6 Goldsworthy, A, 2009 *How Rome Fell: Death of a Superpower*, New Haven.
7 Gosalvez, Emma, 2020 "How Habitat Destruction Enables the Spread of Diseases Like COVID-19," *College of Natural Resources News* (North Carolina State University) https:www.news.nscu.edu Accessed on 8/1/2022.
8 Greer, John Michael, 2005 *How Civilizations Fall: A Theory of Catabolic Collapse*, (Ecoshock).
9 Greger, Michael, 2020 "The Covid-19 Pandemic May Just Be a Dress Rehearsal," Covid-19 Series. https://www.NutritionFacts.org July 2020, 50 Cov. Accessed on 10/25/2020.
10 Hagens, Nate, 2022 "*The Great Simplification*," https://www.thegreatsimplification.com .

11 Hagens, NJ and White, D.J, 2021 *Reality Blind: Integrating the Systems Science Underpinning Our Collective Futures,* (Amazon) p. 19.
12 Hallett, Steve and Wright, John, 2011 *Life Without Oil: Why We Must Shift to a New Energy Future* (Prometheus).
13 Harper, Kyle, 2017 *The Fate of Rome: Climate, Disease & the End of an Empire* (Princeton University Press).
14 Harper, Kyle, 2021 *Plagues upon the Earth: Disease and the Course of Human History* (Princeton University Press).
15 Harvard T.H. Chan (Harvard School of Public Health) 2020 "*Nutrition and Immunity*" https://www.hsph.harvard.edu/nutritionsource/nutrition-and-immunity/ Accessed 8/1/2022.
16 Jonas, K.E and Patel, N, 2008 "Global Trends in Emerging Infectious Diseases," *Nature* 45: pp. 990–994.
17 Kwon, Jennie and Powderly, William, 2021 "The Post-Antibiotic Era is Here," *Science*, https://doi.org/10.1126/science.abl5997
18 "Leo and Majorian Demand the Preservation of Rome's Buildings, AD 458 in Lewis, Jon, Ed. 2003, *The Mammoth Book of Eyewitness Rome* (Carroll and Graff), pp .447–449.
19 McCormick, Michael et al. 2012 "Climate Change During and After the Roman Empire: Reconstructing the Past from Scientific and Historical Evidence," *Journal of Interdisciplinary History* 43: pp. 169–220, pp. 190–191.
20 McCormick, Michael, 2011 Review: *History's Changing Climate: Climate Science, Genomics, and the Consilient Approach to Interdisciplinary History*, 42(2): pp. 251–273.
21 Morand Serge and Walther, Bruno, et al. 2020 "The Accelerated Infectious Disease Risk in the Anthropocene: More Outbreaks and Wider Global Spread," *BioRxiv* https://doi.org/10.1101/2020.04.20.049866.
22 Pathak, Neha and McKinney, Amanda, 2021 "Planetary Health, Climate Change, and Lifestyle Medicine: Threats and Opportunities," *American Journal of Lifestyle Medicine* https://journals.sagepub.com Accessed on 4/22/2021.
23 *Planetary Health Alliance* 2022 https://www.planetaryhealthalliance.org/planetary-health Accessed on 8/1/22.
24 Purdon, Liam 2021 "*Codex Exoniensis, fols. 123b-124b:* An Old English Poetic Romano-British Encomium," *Quidditas, Online* Journal of the Rocky Mountain Medieval and Renaissance Society.
25 Quammen, David, 2013 *Spillover: Animal Infections and the Next Human Pandemic* (New York: W.W. Norton, 2013).
26 *Rome: Power and Glory*, 2010 Parts I-VI https://www.imdb.com

27 Roxby, Philippa, 2022 "Millions are Dying from Drug-resistant Infections, Global Report Says," *BBC World* https://www.bbc.com/news/health-60058120 Accessed on 1/20/2022.
28 Sundstrom, Shana 2017 "The Adaptive Cycle: More than a Metaphor," *Ecological Complexity*, 34. https://doi.org/10.1016/j.ecocom.2019.100767
29 Tainter, Joseph, 1988 *The Collapse of Complex Societies* (Cambridge), inclusive.
30 Turow-Paul, Eve and Egan, Sophie, 2021 "Dietary Guidelines for Americans: There Are No Healthy People on an Unhealthy Planet," https://foodtank.com/news/2021/01/dietary-guidelines-for-americans-there-are-no-healthy-people-on-an-unhealthy-planet/#:~:text=Dietary%20Guidelines%20Americans%3A%20There%20Are%20No%20Healthy,holidays%20When%20the%20recommendations%20are%20anachronistic%20on%20arrival. Accessed on 1/24/2021.

Index

M. Orsag et al., *Interdisciplinary Insights from the Plague of Cyprian*, https://doi.org/10.1007/978-3-031-26094-0

Printed in the USA
CPSIA information can be obtained
at www.ICGtesting.com
LVHW021448221123
764620LV00001B/15

9 783031 260933